RECYCLING IN THE
GARDEN

RECYCLING IN THE
GARDEN

ANGELA YOUNGMAN

WHITE OWL

AN IMPRINT OF PEN & SWORD BOOKS LTD.
YORKSHIRE – PHILADELPHIA

First published in Great Britain in 2022 by
White Owl
An imprint of
Pen & Sword Books Ltd
Yorkshire - Philadelphia

ISBN 978 1 39900 183 0

Typeset in 11/14 pts Cormorant Infant
by SJmagic DESIGN SERVICES, India.
Printed and bound in the UK by CPI Group (UK) Ltd., Croydon. CR0 4YY.

Pen & Sword Books Ltd incorporates the imprints of Pen & Sword Books Archaeology, Atlas, Aviation, Battleground, Discovery, Family History, History, Maritime, Military, Naval, Politics, Railways, Select, Transport, True Crime, Fiction, Frontline Books, Leo Cooper, Praetorian Press, Seaforth Publishing, Wharncliffe and White Owl.

For a complete list of Pen & Sword titles please contact

PEN & SWORD BOOKS LIMITED
47 Church Street, Barnsley, South Yorkshire, S70 2AS, England
E-mail: enquiries@pen-and-sword.co.uk
Website: www.pen-and-sword.co.uk

or

PEN AND SWORD BOOKS
1950 Lawrence Rd, Havertown, PA 19083, USA
E-mail: Uspen-and-sword@casematepublishers.com
Website: www.penandswordbooks.com

Contents

Introduction   6

Natural Recycling   9

Learning from the Past   17

Water Recycling   29

Energy Recycling   42

Green Roofs and Green Walls   57

Recycling Materials   68

Repair, Reuse, Recycle   114

Upcycling and Case Studies   127

Conclusion   154

Resources   156

Index   157

Introduction

The world is changing. People are becoming more aware of the environment and their impact on it. Over the past decades we have become an increasingly consumerist society. From a world in which recycling was common, single use became the norm. A throwaway society has developed, covering everything from food to electronics. New was always regarded as being better than old, and throwing away rather than repair has become a normal part of life. People are buying more and more new products, disposing of ever greater quantities of unwanted or broken merchandise, and using raw materials at a rate never before experienced. This situation has become unsustainable, with environmentalists and scientists drawing attention to the finite nature of many natural resources, together with the impact of climate change and its implications for the environment, lifestyle and gardens.

The burning of fossil fuels to provide power and heat has led to increasingly high levels of greenhouse gases trapping the sun's energy and making the earth much hotter. As a result, extreme weather conditions such as heavy snowfalls, storms, floods, heatwaves and droughts are becoming more and more common.

Demand for all types of energy has rocketed worldwide. In 2019, Statistica.com reported that global demand for crude oil amounted to 100.1 billion barrels daily, while the discovery rate of new oil resources amounted to just 12.2 billion barrels annually. Compare this to the situation in 1955, when the world was using 4 billion barrels of oil annually and discovering around 30 billion barrels of new crude oil resources. Such excessive use is not limited to oil – a similar situation can be found with regard to all types of resources.

Peat is a typical example of unsustainable extraction of raw materials. It is one of the most popular growing mediums used by gardeners, but it is using a material that can never be replaced. The Plantlife website states:

> Peat is plant material which is partially decomposed and has accumulated in waterlogged conditions. Peat lands include moors, bogs and fens, as well as some farmed land. Peat bogs grow slowly, accumulating around 0.5 to 1mm of peat each year, and the water prevents the plants from decomposing. Many areas of UK peat bog have been accumulating gradually for as much as 10,000 years,

Peat, a finite material being dug from the soil.

and can be up to 10m deep. Due to its slow accumulation, peat is often classified as a fossil fuel.

Commercial peat extraction together with increasing use of drainage to provide agricultural land has destroyed the unique eco-systems within peat bogs that existed for centuries and enabled them to act as carbon dioxide reservoirs, thus significantly contributing to climate change. The National Trust points out:

> Peat holds more carbon than the combined forests of Britain, France and Germany. It holds up to 20 times its own weight in water. Dry peat releases carbon dioxide and is one of the biggest sources of greenhouse gas.

Gardening is by far one of the biggest users of peat. Plantlife say that amateur gardeners are responsible for the use of 3 billion litres of peat every year, accounting for 69% of peat compost used in the UK. A total of 32% of that peat is sourced from the UK, 60% from Ireland and 8% from Europe. Commercial extraction can remove more than 500 years of 'growth' within just one year. It also adds to the problem of climate change because when peat is mixed with soil and exposed to aerobic conditions, it begins to decompose, releasing carbon dioxide into the atmosphere and thus further contributing to the greenhouse effect.

Garden Organic is one of many gardening organisations that campaigns against the use of peat in potting composts, pointing out that 95% of the UK's peat bogs have already been destroyed and that much of the peat-based potting mixes available are now being sourced from worldwide sources. Peat-free composts are available, and can prove much more successful than peat-based variations. The Royal Horticultural Society has shown that peat-free gardening is viable, as its gardens are now 97% peat free and it is committed to reducing peat use wherever possible.

The campaign against peat use is just one aspect of the reawakening of the waste not, want not mentality, which had once been a dominant feature of society. Recycling and repair has been regarded as an important facet of human life for many centuries, with earlier societies throwing away as little as possible. In the modern world, recycling is increasingly important since it reduces the amount of waste being sent to landfill and reduces problems caused by pollution. By opting to recycle materials it reduces the need for extracting and processing raw materials, as well as reducing greenhouse gas emissions since less energy is required in the production process. In this changing world, gardeners are very much at the forefront of the recycling and reuse movement, which covers every aspect of gardening techniques and practices including water, energy and materials.

Recycling and reusing is good for the environment and makes financial sense, saving money and time. Think before you buy or throw something away – is it worth it? Is there too much packaging involved? Is there a greener alternative? Can items be recycled or upcycled into something totally different?

RECYCLING IN THE GARDEN

Natural Recycling

Natural recycling has always been a key element within every garden, although perhaps not automatically recognised as such. Often described as sustainable gardening, it results in the creation of an environmentally friendly area in which natural predators thrive and soils are naturally replenished.

Tailoring gardens to suit the environment, using plants that match the soil and climate, is crucial to the success of natural recycling sustainable practices. Instead of trying to impose a particular garden style such as a dry, Mediterranean or plant ericaceous species in areas with unsuitable soil, a sustainable garden could be more easily created by matching the plants to the soil. This avoids having to spend a lot of time and effort trying to alter the composition of the soil and the environment, using materials such as gravel and peat that may have been transported long distances or that involve the use of non-renewable resources.

An environmentally friendly garden incorporating natural materials.

When making alterations to a garden – especially if you have just moved into a new property – take time to study the environment. Is the soil naturally dry or wet? Where are the boggy or wet areas? If you have a natural pond, ditch or stream running through the garden, don't fill it in – keep it well maintained and remove any debris. Ponds and streams are nature's own way of moving water from the land and if they are blocked, flooding will result whenever there is heavy rain. Rain gardens have experienced a growth in popularity, comprising grassed-over hollows that are allowed to flood or stay marshy and boggy to cope with heavy rainfall, with the option of draining and recycling the surplus water using porous surfaces and underground tanks.

At Beehive Farm in Wales every attempt is made to recycle and leave no impact on the environment. Owner Matthew Watkinson has sought to create a natural edible landscape in which plants match the location. Raspberries are allowed to grow where they like and experiments are under way with alternative crops such as nettles and wild garlic, which are extremely nutritious and useful. His gardening techniques are simple – he purchased a scythe and now regards this as his most essential tool, cutting down raspberries, bracken and nettles as necessary.

Recycling the natural materials created within the garden is simple. Rather than collecting up all the autumn leaves, natural recyclers leave some on the ground, especially in flower beds or shady areas. Worms and other insects steadily move the

Gravel mulch in use to preserve water in a front garden. (Angela Youngman)

Collecting seeds from marigold plants.

leaves underground over the winter, composting them naturally into the soil. This helps create a beneficial environment for insects and small mammals like hedgehogs that like to hibernate amid piles of leaves. Creating compost from brown and green waste sourced around the garden and kitchen is an automatic gardening task.

The hollow stems produced by hogweed and nettles are ideal for overwintering bumblebees and other insects. Collecting up bundles of such stems and putting them together in a sheltered place will create bee hotels. Over time, they will naturally degrade, and can be simply replaced as necessary.

Mulching regularly will minimise water usage since it provides a protective covering for the soil, deterring weed germination and weed growth by blocking out light and retaining moisture. There are a variety of mulches that can be used, many of which involve using recycled materials such as newspaper, cardboard, plastic sacks, bark or wood chips. Hoeing also helps since it removes weeds and reduces competition for any available water. It keeps the surface friable and loose, enabling rain water to pass through the surface to reach plant roots. Even placing stones around the base of vulnerable plants will help trap moisture and keep roots cool throughout the summer. Gravel is one of the most frequently used mulches.

The life cycle of plants offers many natural recycling opportunities. Annuals, biannuals and perennials can be grown from seed, created naturally at the end of their flowering period. Collecting such seed was an essential task in days gone by as this was the only way that gardeners could guarantee plants for the following year. For many plants this is still an option for modern gardeners, unless the parent plants are F1 hybrids. Seed from F1 hybrids do not reproduce well, as they require special cross-pollination. Almost any other plant, especially wild flowers, can be grown from seed. Choose a dry, sunny day to collect the seed. Gently shake the flower heads into envelopes or small brown paper bags. Label with the name and year, then keep in a dry, dark place such as a container with a lid that can be fastened securely. Take care not to squash the seeds within the container. Sow the seeds as normal the following year.

Keeping a lawn in perfect condition can involve lots of fertilisers, conditioners and regular mowing. Cutting less frequently and opting for a longer grass height immediately results in lower energy use, as well as creating grass that is more resistant to periods of drought. An even better option is to use a mulching mower, which automatically turns grass clippings into tiny pieces, returning them instantly to the soil to act as a natural mulch and soil conditioner. It avoids any need to collect up grass cuttings after each mowing, and keeps the lawn in much better condition. Watering lawns during the summer should be avoided as it wastes water. Even if grass looks dead, it will revive quickly once the rain comes. Grass can survive for several months without water before dying completely.

Green manures recycle nutrition back into the soil as a natural soil conditioner/ fertiliser. Using a green manure will also decrease weed growth as it avoids having expanses of empty soil. They are particularly useful in the vegetable garden. A green manure is basically a crop that is grown for a short period of time such as phacelia, winter vetch, alfalfa and mustard. Some crops such as mustard are quick growing and mature rapidly, making them ideal as an intercrop or just as a way to fill empty soil for a few weeks. Others such as lucerne require a long growing season. Choosing a green manure depends partly on timescale and the type of nutrition required. Good nitrogen-fixing plants are lucerne, clover, bean, lupin and tares, whereas rye, buckwheat, phacelia and mustard do not fix nitrogen. Green manures must be dug into the soil before they start to flower and set seed. Once dug in, the plants decompose quickly, releasing nutrients that can immediately be utilised by the next crop of plants to be grown on the site. Large plants should be mowed first before digging into the soil, as this will aid decomposition. After cutting down the plants, leave the green growth on the soil for a few days to wilt. When digging in green manures, take care not to bury the material too deep, ideally aiming for a depth of around 15cm (6in).

Creating natural fertilisers from nettles, comfrey and manure is a frequent option, as are other natural compounds. Listening to a talk by Peter Beales, of Peter Beales Roses, a group of Norfolk gardeners were fascinated to discover that he used a 50-50 semi-skimmed milk-water mix as a spray for blackspot on roses.

Tree management offers countless opportunities for natural recyclers. Wildlife welcome the creation of small log piles, especially in shady areas as the decomposing

Using mustard as a green manure.

logs provide shelter and food. Faced with the need to recycle turves and logs while redesigning a garden, landscaper Simon Smith created a 'beetle bank'. Designed as an ecological, sustainable solution, the bank was constructed by placing a line of turves on the ground. More turves were placed on top in an interconnecting brick format, until it measured about 400mm high. Logs were then placed across and behind the turf wall. As the logs and turf decompose, it forms a dark, damp home for beetles and other insects.

Other logs can be placed in a stumpery, sawn up to provide stepping stones, or placed to create path supports especially within woods. Logs and branches that cannot otherwise be recycled can be put through a wood chipper and turned into chippings ideal for use on paths. All these materials will eventually rot down, encouraging the development of beneficial fungi and resulting in natural recycling. As Nick Fraser, head gardener at Nunnington Hall in North Yorkshire, points out, 'It is a closed loop system.'

Historic gardens such as Arundel are very aware of the need to use natural resources, whether it is by creating tepees from pollarded Paulownia tormentosa trees or just collecting the fallen leaves to produce leaf mould. Hazel is harvested to provide runner bean wigwams and archways for climbing sweet peas. When a holm oak fell in high winds, it was quickly recycled into numerous benches placed within the garden,

A plastic bottle bird feeder hanging outside a window.

and timber for the newly built thatched roundhouse and boathouse were obtained from the estate woodlands.

An important aspect of natural gardening styles is the need to be environmentally friendly, and consider the needs of wildlife. Recycling can help wildlife and attract greater numbers of useful insects, birds and mammals to the garden. Creating bird feeders from plastic bottles and filling tubs with fat balls are popular tasks, especially with children.

Creating a safe place for birds to access bathing and drinking water all year round is essential, especially in summer when water is scarce. An old bin lid or plant tray can easily be reused to make a simple bird bath since all that is needed is a shallow, watertight bowl with sloping sides to provide birds with a safe place to stand and drink, sited on a plinth that can be made from old bricks or rocks.

Another popular recycling technique to help garden wildlife is the creation of bug hotels. These provide homes for a range of insects, such as ladybirds, beetles and solitary bees, all essential for ensuring propagation and dealing with aphids around the garden. These can be as simple or as complex as required. At its simplest, it is a collection of hollow stems pushed inside a plant pot to provide suitable shelter.

Above: *Bamboo sticks tied together form a simple insect hotel.*

Below: *An insect and butterfly bug hotel created out of pallets.*

More complex ones can be made from a range of open-fronted containers such as old tyres, biscuit tins, storage drawers, deep picture frames and pallets, or they can be custom-built using spare pieces of wood and old tiles. Larger containers can be divided up by using smaller containers inside such items as bottles or plant pots. Layer materials such as hollow bamboo canes, pieces of hosepipe and small tube firmly inside the frame. Fill in any surplus space around the hollow tubes with small branches, twigs, bark, wood chips, dried leaves and pinecones as these can provide additional housing for various insects such as ladybirds and beetles. Corrugated cardboard makes a popular home for lacewings but should be placed towards the centre of the bug home so that the cardboard is sheltered from any water incursions by other resistant materials. Place the bug home on a raised surface so that it has some protection from any flooding that may occur over winter, and make sure that the roof is water resistant. Old terracotta roof tiles are ideal for this purpose.

Bear in mind that small logs also make good insect homes. Beetles and spiders enjoy the shelter provided by the bark, while drilling holes into the circular centre of the trunk will provide homes for bees and ladybirds. The strong bark surface helps to protect the sheltering insects from the winter weather.

It is also worth considering placing some old roof tiles and rocks on the ground to provide dark crevices suitable for frogs and toads – after all, they love munching slugs, so they make a very welcome addition to any garden.

Learning from the Past

Throughout history, gardeners have proved to be extremely innovative and have avoided waste wherever possible. Many of the recycling techniques used by gardeners over the centuries are proving to be just as effective today or have provided inspiration for answers to modern-day problems. Visiting living history museums such as Ironbridge, the Black Country Museum, Beamish and Eden Camp can be extremely informative as you can see traditional techniques and garden styles in action. There are small Victorian gardens attached to various houses at Ironbridge, Beamish and the Black Country Museum, likewise at Gressenhall Farm and Workhouse museum. Recreated Second World War gardens complete with Dig for Victory allotments and air raid shelters can be seen at Beamish and Eden Camp.

A hand pump and covered well within a wartime country garden. (Angela Youngman)

The realities of wartime gardening are very evident at Eden Camp in Yorkshire, which was a Second World War prisoner of war camp and home to more than 1,000 captured soldiers, mainly from Eastern Europe. The PoWs were encouraged to create gardens throughout the camp, and went out to work on local farms and orchards. Their gardens within the camp were created on a shoestring, with few resources available. In order to demonstrate the wartime gardening techniques available to the prisoners as well as home front gardeners throughout the country, an allotment has been created on site.

Water storage was the first priority. During the war, there was always a risk that mains water would be disrupted as a result of bombing and all households constantly faced the possibility that there would be none available. Prisoners of war and domestic gardeners were not allowed to use mains tap water. Setting up a network of water butts to store rain water was essential, as was using grey water from around the buildings, although the amount of washing water was always limited since people were only allowed 4 inches in the bath. Creating compost heaps was an essential part of gardening, while everyone learned to grow vegetables, with 'how to' leaflets and books being made widely available. Among the key techniques replicated on the allotment were the use of recycled wood stakes to deter pigeons, and the creation of a scarecrow to frighten birds away, as well as rats.

For centuries, manure from both humans and animals was collected and spread on the ground as natural fertiliser. Most houses had 'bucket toilets' that had to be emptied regularly, with some of the contents being used on the land.

Interest in compost toilets is beginning to re-emerge. They are particularly useful on allotments or remote areas that attract visitors, such as woodlands. There are several benefits, particularly as there is no need for mains sewage since solid waste is dealt with on site. They save water, and allow organic materials to be returned to the soil. No chemical bleaches or cleaners are required.

Compost toilets are basically waterless toilets. The simplest form is a bucket system. The bucket is placed under the toilet seat and the contents kept covered at all times with straw or sawdust. It has to be emptied regularly onto an outdoor compost pile.

Much more popular are the self-contained systems. When used, the contents fall into a chamber, where they remain to decompose. There are usually two chambers – one used for a year before the seat is removed and placed on the second chamber. This allows the contents of the first chamber to decompose before being emptied twelve months later. Adding a soak (a handful of straw or sawdust) each time the toilet is used avoids any problems of unpleasant odours since it prevents methane, hydrogen sulphide and excess nitrogen in the form of ammonia being produced. Daily maintenance is essential. Compost toilets have to be checked regularly to ensure that no problems are developing and that buckets of straw or sawdust are kept topped up. The Centre for Alternative Technology recommend that urine diverters are incorporated into the toilet design as this can be diluted with water to use as a fertiliser on non-food plants. It is generally recommended that the contents of compost toilets should be used on trees and bushes rather than on the vegetable garden.

Farmers and home gardeners have always used manure from livestock and poultry on the soil as the manure contains many useful elements, including nutrients, organic matter and fibre.

In *The Wartime Kitchen Garden*, Jennifer Davies recounts the story of Mike Benson from Liverpool. As a boy growing up in wartime, he was frequently sent out on his bike with canvas bags attached to his handlebars. His destination was the local pub. When the dray horses arrived with the beer delivery, they would often relieve themselves on the road. Mike was expected to scoop up the manure, and take it home for use in the garden.

A slightly different slant on the use of manure was also recorded in *Wartime Kitchen Garden*. Land Army girl Mary Smith worked on a farm and her dungarees tended to became very stiff due to deposits of pig swill, cow, pig and horse manure. She sent the dungarees home to her mother for washing. Her mother was apparently extremely pleased since soaking the dungarees in water before washing them resulted in a very nutritious liquid fertiliser that was perfect for tomatoes!

Nunnington Hall is a large historic estate in North Yorkshire, which is now in the care of the National Trust. The extensive gardens are managed using a range of traditional techniques that are designed to be as sustainable as possible. One of those techniques links two historic agricultural sectors from the area – apple growing and sheep farming – and has resulted in extremely high crop levels.

The Nunnington Hall orchard possesses more than twenty-five varieties of apple, including local specialities such as Dog's Snout, Yorkshire Cockpit, Ribston Pippin and Yorkshire Beauty. The orchard is located within a wildflower meadow and while the trees require fertiliser to be productive, the wildflower areas require low nutrition. This is achieved using traditional recycling and cultivation methods found within the region. Wool fleeces are used as mulch around the trees. Nick Fraser, head gardener, explains:

> It is a practice which has been used for centuries. The area has always been a big apple growing region, as well as sheep farming. Good fleeces were sent away to market, and farmers were left with 'bag ends' – the bits of fleece covered in manure that were difficult to clean. These bag ends were placed around the base of fruit trees. They decompose naturally and allow natural nutrients from the manure to pass into the ground, keeping the fruit trees healthy. The trees are fertilised but not the wild flower area. It is very effective.

Scarecrows scaring birds from crops.

There are also added environmental benefits. Apart from acting as a mulch and soil conditioner, the fleeces provide insulation and a home for beneficial insects, small mammals and worms. Mice and bumblebees like to nest underneath the fleeces.

Just like gardeners over the centuries at Nunnington, the modern gardening team aims to recycle and reuse as much as possible. Hazel poles are turned into plant labels, first shaving off a small section so that plant names can be written on clearly. Any spare bits of wood, even damaged or unwanted furniture from the house, are cut down and upcycled, such as when the gardeners were given a lot of wooden boxes that had originally formed part of a stage used for theatre performances in the hall. These boxes were no longer needed for their original purpose and were quickly altered to become bird boxes placed around the estate.

Scarecrows are a traditional way of scaring birds away from newly planted crops, with numerous variations being made. Covering a potato with feathers and leaving it on a stick to scare the birds away from newly planted seeds is possibly one of the most unusual forms of bird scarer used by gardeners in the past.

Other gardeners have begun to use tools more commonly used in previous centuries. At Beeview Farm in Pembrokeshire, Matthew Watkinson has revitalised traditional practices by using a scythe to cut bracken, rush and grass for animal bedding, as well as incorporating natural foods like nettle and wild garlic into the family diet. Having

RECYCLING IN THE GARDEN

initially expected to use it occasionally, using a scythe has proved so successful that it has become one of his most popular tools.

Dealing with pests such as wasps has always been a problem for gardeners, who improvised wasp traps from jam jars. Gardeners half filled a jar with sugared water. The top of the jar was covered with thick paper and tied in place. Holes big enough for wasps to crawl through were cut into the top. The wasps were attracted by the sweet smell, entered the jar and drowned. This technique continues to be used by many gardeners today.

During the eighteenth and nineteenth centuries, the activities of plant hunters travelling to far-off lands transformed gardens as they brought back rare and unusual plants such as varieties of rhododendrons, ferns and lilies. Sealed, protective containers known as Wardian cases were used to transport many of these plants, especially cuttings and seedlings as the containers recycled moisture into the plants during journeys. These cases were a forerunner of the modern terrarium. Modern gardeners often use large jars or small cloches, which can be sealed up to recreate a version of a Wardian case as a way of displaying several small plants. When creating such a container, it is important to ensure that it can be cleaned periodically.

Revival of the Victorian stumpery

Disposing of tree stumps can be difficult due to their size and weight. Traditionally, stumps were generally sawn up and burned on the fire. In Victorian times, a new and innovative use was found for tree stumps, which effectively allowed them to be upcycled into a new style of gardening by creating a stumpery. These stumperies became an extremely popular garden feature and were incorporated into countless gardens nationwide.

The concept of the stumpery emerged due to the Victorian fascination with ferns. In the early Victorian period, gardeners keen to add ferns to their gardens stripped the countryside and imported still more. Having collected their ferns, the Victorian collectors wanted to show them off. This led to the development of extremely creative ways of displaying them within a domestic garden, especially those belonging to aristocratic houses. Tree stumps became the accepted method of display because when the roots were carefully placed they could act almost like a rockery for ferns, providing moist planting areas in the shade of the roots.

Biddulph Grange in Staffordshire was the site of the first ever stumpery. Designed by James Bateman, it involved piling up tree roots and turning them into a fern-loving garden area. In October 1856, the *Gardeners' Chronicle* noted that a 'rustic root garden' had been created at Biddulph Grange. Two months later, the periodical took a more in-depth look at this new style of gardening, describing it as 'the Stumpery' for the first time.

A massive stump surrounded by lush greenery rears up against stone walls in the atmospheric stumpery at Arundel Castle. (Martin Duncan, Arundel Castle)

The magazine noted:

> The root garden, or as it is here called, for want of a better term, 'the Stumpery,' consists of a very picturesque assemblage of old roots or rugged stems and stumps of trees – chiefly the latter – piled to a height of 8 or 10 feet on either side of a winding and rapidly descending walk. They are so irregularly arranged as to jut forward in the boldest prominence, and even to be united into a rustic arch in some parts; while in others they recede far enough to allow room at their base for little gatherings of choice herbaceous plants, bulbs or miniature shrubs.
>
> Mr Bateman, has been singularly fortunate in procuring a quantity of the most gnarled, contorted and varied masses of wood imaginable for this purpose, and they are joined together and disposed with consummate art. The blocks being all of Oak too, they are likely to be very durable. Over considerable portions of the whole, masses of Ivy, Virginian Creeper, Cotoneaster and other trailing plants scramble about in the wildest manner. And the interstices, as well as the open spaces now and then occurring at the base, are all used for the reception of some characteristic and interesting plant or group. For example, near the entrance to this region, the Hellebores, which are among the earliest of the winter-flowering

RECYCLING IN THE GARDEN

plants, are clustered in great variety. Then follow the Anemones, Epimediums, Scillas, Dog tooth Violets, Lilies of the Valley, each kind receiving the precise amount of sunlight or shade which is desirable for it, and all being intermingled with Gaultherias, Pernettyas, Cotoneasters, Savins and such other dwarf evergreens as serve to produce a sufficiency of green clothing at all seasons of the year. Even the rarer hardy Orchises and the Cypripediums, have an appropriate corner assigned to them, and seem quite at home in it.

These descriptions of The Stumpery immediately aroused considerable interest among fern loving gardeners. Within a short time, a nearby household, Arley Hall in Cheshire, had followed suit. Named 'The Root tree', the Arley Hall stumpery was said to resemble a miniature mountain landscape complete with stumps, grotto and pools, all planted up with a large collection of ferns. Many others were created elsewhere, but due to their transient nature with the stumps being allowed to slowly decompose, along with changes in gardening styles, most of the original Victorian stumperies have disappeared over the years.

One of the few exceptions is at Ickworth in Suffolk, the home of the Marquess of Bristol, where a stumpery was created in the Italian Garden. Now in the care of the National Trust, Ickworth's stumpery has been maintained and extended. The great storm of 1987 provided a vast amount of raw materials due to the number of trees toppled by the wind. As a result, an extensive additional stumpery has been created that includes features such as a dragon's lair complete with a clutch of dragon eggs. Jack Linfield, Ickworth's head gardener, has commented that the stumpery resembles Alfred Rackham's fairyland illustrations:

All the stumps in the new section were mostly oak trees blown up during the Dig for Victory phase. They are as hard as rock, and very durable. We also went through the woods with a digger and dug up the roots of other trees that had died, prising them out of the ground.

The stumps are very gnarled, blending in and complementing the shapes and gnarled appearance of the trees around them. We want to showcase the roots, and create lush planting around rather than in the stumps. It is a mix of greens, leafy foliage and flowers with the odd splash of colour.

Typical plants used in the lush scheme include Galium Odoratum (woodruff), euonymus kewensis, dicentra, cobra lily, hellebores, wood spurge, ruscus and scopolia. Some unusual techniques have been used to maximise the lush ambiance: noticing a honeysuckle arching between trees, it was encouraged to continue spreading in this fashion by placing a rope between them. Another unusual element is the use of the Mind Your Own Business plant, which spreads very fast. Jack Linfield has used this to cover some of the drier shady areas as it holds in moisture, as well as placing it on a stumpery wall constructed to highlight a pathway. Each wall contains approximately ten stumps, which were placed in position using a mini digger and three gardeners to manoeuvre the stumps into place.

A different approach was taken at Batsford Arboretum in Gloucestershire. In 2000, the gardeners at Batsford combined a bog garden and stumpery to create a 'swampery'. Partially submerged large tree stumps were used to grow primulas and foliage plants such as gunnera, skunk cabbage and tree ferns.

One of the most influential modern stumpery gardens was created at Prince Charles's home at Highgrove. It comprises 180 sweet chestnut tree roots sourced from across the estate. The immediate effect of a new stumpery does look somewhat unattractive: it is reported that when Prince Philip saw the initial project he promptly asked, 'When are you going to set fire to this lot?' Since then, the stumpery has been transformed into an extremely pleasant and eye-catching area in which the stumps are planted with hostas, ferns, euphorbias and hellebores. Some of the stumps are held together with steel in order to create large arches over pathways. The stumpery is now one of the most popular sections of the garden whenever it is open to the public.

In 2011, Burnby Hall Gardens near York began work on a stumpery inspired by Highgrove's version. According to staff, it has now become one of the most popular of all the themed gardens on the estate. Stumperies have reappeared in designer gardens at the RHS Chelsea.

At Biddulph Grange the stumpery remains a centre of attention to visitors. Head gardener Paul Walton says:

What makes Biddulph's stumpery stand out is the location [situated between the Chinese garden, East Terrace and Dahlia Walk]. It is surrounded by very formal areas and is so well hidden that you only see it at the last minute and with some of the bankside being three to four metres high, it really does stand out. We don't overplant this area as I feel the focus needs to be stumps, so we have mainly ferns and some spring bulbs in this area.

At Glacier Gardens in Juneau, Alaska, Steve and Cindy Bowhay have taken the concept of a stumpery to new heights. Located amid the lush south Alaskan temperate rainforest habitat, they take visitors on tours through old-growth rainforests as well as their unusual stumpery gardens involving both the trunks and roots of old trees. Steve says,

I use trees that have blown over with intact root systems, I trim and clean out the roots before standing them up. I dig a deep hole about 10 feet using an excavator, then lift the stump with chains so that the trunk is facing down and insert it into the hole. Once secure, I fasten fish net over the top of the roots. I place live

RECYCLING IN THE GARDEN

A colourful transformation of tree trunks and roots at Glacier Gardens in Juneau, Alaska. (Steve Bowhay)

moss on the net facing down and cover the roots of the moss with potting soil, then I plant the flowers.

Apart from their long-term attractiveness as a growing area, a stumpery is good for wildlife since it provides the ideal habitat for stag beetles, toads and small mammals. Once in place, the stumps are left to decay.

Stumperies are easy to create, and just require a suitable site, surplus wood and some plants. They can be as grand or as small as you want. It can involve a large quantity of tree stumps or just a handful of logs. Norwich-based designer Rajul Shah was inspired to create a small stumpery in her front garden when she saw some stumps being taken to the dump. It took just a few hours for her to dig over the chosen site under a cherry tree, put the roots in place and add a selection of bulbs and shade-loving plants.

If not available from trees being cut down within your own garden, tree stumps can often be obtained from tree surgeons, building sites and managed woodlands. It is a matter of asking around. The amount of stumps required depends on the size of your project. Smaller projects may require just a few logs or large branches that can be positioned carefully together, interlocking to create pockets suitable for planting. Such clumps can look very pretty around ponds or cascades, as well as in shady areas.

When creating a stumpery, hardwood stumps such as oak and beech are particularly good as they rot slowly. It is important to fork over and remove any

A small stumpery log feature created in Rajul Shah's Norfolk garden. (Rajul Shah)

perennial weeds. Old stumps need to be dug into the soil at least 3ft deep in order to stay safely in place. It is very important that the stumps and logs are secure, and do not rock or slip. Always turn the stumps so that the trunk is in the hole, and the roots are sticking up into the air. In general, at least two or three stumps can be linked together in one planting area. Position them so that they form a group, leaning or resting on each other. The aim is to create a natural look. Place logs, or intertwined sections of old wood to create a loose boundary around the stumpery, which will help develop shade levels and ecological corridors for the wildlife. Add compost into crevices around the root to provide a good base for plants. As the wood rots down, it will automatically replenish the nutritional quality of the adjoining soil. Apart from ferns, good plants to use in stumperies include wood anemone, dog's tooth violet and lily of the valley, as well as foxglove, primroses and bulbs including bluebells and snowdrops.

Bark chips scattered around the surrounding area will enhance the woodland effect. Mosses, lichen and fungi can be encouraged to grow on the stumps by covering the logs with a layer of natural yoghurt.

Once created, a stumpery is a low-maintenance form of gardening. It will develop naturally and provide a lot of interest within the garden. Paul Walton, head gardener at Biddulph Grange, says,

There was some restoration work done during the 1990s to replace some of the rotten original stumps and since then more stumps have been added but not replaced. The stumps we use are oak as once all the small roots have rotted away you are left with this fabulous shape. When we have stumps delivered, we tend to leave them a couple of years to allow them to grow moss/lichens, so when we add them to the Stumpery they blend in.

Arundel Castle case study

Martin Duncan, head gardener at Arundel Castle, outlines how he created a new stumpery within the grounds:

Following the devastating hurricane of 1987, which brought down and uplifted endless ancient trees throughout the southern counties, the castle estate foresters had years of clearing up, resulting in the ancient stumps being left where they lay. This is what gave me the idea in 2014 to use these magnificent stumps, so I chose mainly oak, yew and sweet chestnut stumps. Some of the yew trees were at least 400 years old when they fell. The foresters sourced the best from the estate woods, high on the South Down hills here in West Sussex, and with their help they were brought into the gardens for me to design a stumpery garden.

I initially drew a design concept for a stumpery garden and presented it to the Duke and Duchess of Norfolk. I wanted to create something quite unique for them and so the first year we kept it small and by the garden wall. This ultimately resulted in it being extended and over the last few years it has transformed this previously empty part of the gardens into something quite magical.

Once the stumps had been brought into the area I placed them all individually. We dug holes to about half a metre deep to secure them in the ground with the help of the

A path winding its way through the stumpery at Arundel Castle. (Martin Duncan, Arundel Castle)

estate foresters and a contractor with a mini digger. I was able to use my creativity by positioning every stump with the roots upwards to show off their magnificent architectural structures, allowing them to become an art form in their own right. The only exception is the centrepiece, which I placed on a large mound of earth. I then laid the two large stumps in a position that gives the garden its central height and focus.

The aim was to give our visitors the feeling of walking through a magical woodland setting. To achieve this, we placed some plants to grow through the stumps and cascade onto pathways. In spring we use small botanical tulips such as turkestanica, linifolia, sylvestris, pulchella – Persian pearl and clusiana – peppermint stick along with English bluebells, snowdrops and narcissi, echium pininana (tower of jewels) dierama pulcherrimum, known also as angel's fishing rod, euphorbias, ferns and martagon lilies, giving a very different type of stumpery garden.

In 2017, I was delighted to be able to add six liquidambar trees to the stumpery, which give wonderful autumn colour. These trees initially formed part of the wedding decoration for the Duke and Duchess's eldest son's marriage in Arundel Cathedral.

This garden changes with the seasons and yet there is always a great deal of colour and life, not only in the plants, but it's a haven to wildlife such as bees and beetles, not to mention the garden cats, Tilly and Pippin, who often sharpen their claws on the ancient stumps. It seems to be enjoyed by both children and adults alike, some even call it the 'Hobbit Garden' or 'Harry Potter Garden' as it allows them to use their imaginations and disappear into another world!

Water Recycling

Recycling and reducing the amount of water needed within a garden has become a priority for every gardener. Although much of the world is covered by water, fresh water is a limited resource. According to the World Water Commission, only 2.5% of the world's water comprises fresh water suitable for the purposes of drinking and cultivation. Of that 2.5%, two-thirds is locked into glaciers and ice caps. Approximately three-quarters of the remaining useable water is provided through rain, storms and monsoons and is not captured for use. Humanity recaptures less than 0.08% of the available rain and storm water, even when extensive flooding occurs as most of it just drains away into the rivers and oceans.

Climate change is making everyone look much more closely at the amount of water we use. Periods of drought and high temperatures are much more common, as are floods and heavy rain or snow. The sheer extent of the increasingly built-up environment in which we live, together with the number of houses built on flood plains, means that flooding is no longer unusual. Every year rivers experience dangerous water levels and houses and businesses are flooded. More houses and more people result in an increasing demand for water. According to the UK Environmental Agency, each person in the UK uses an average of around 150 litres of drinking water every day with most of it used for cleaning, flushing toilets and watering lawns and flower beds.

As a result, the cost of fresh water is rising, and there are increasing environmental problems being incurred due to the practice of

Water – a scarce commodity. (Angela Youngman)

Plastic bottles turned into watering aids in a vegetable patch. (Angela Youngman)

pumping water over long distances, then disposing of water after it has been used. Summer water shortages and hosepipe bans are becoming more common. Using a sprinkler or hosepipe attached to mains water is now an expensive option for gardeners. A high-pressure rotating sprinkler designed to water lawns and flower beds can use up to 1,000 litres per hour. Paving over drives and gardens to provide parking for vehicles or extending buildings in urban areas means that instead of rain water draining directly into the ground, it runs into the drainage system on roads. As a result, during periods of heavy rain, often known as storm surges, flooding occurs since the drainage systems are unable to cope.

There are measures that can be taken to conserve water usage within the garden. Use natural watering methods such as scooping out the soil beside vulnerable plants to create a dish effect to retain rain water, thus ensuring that more water reaches the roots rather than spreads across the earth. Hoeing regularly will create a friable soil surface, encouraging the water to drain into the soil. Mulching around plants will decrease weed growth and the amount of evaporation that takes place. Plastic watering bottles can be placed in hanging baskets or beside vulnerable plants in

RECYCLING IN THE GARDEN

A porous pipe system installed around a flower bed.

the garden to allow easy topping up when necessary. Water-retaining gel can be beneficial within hanging baskets as it swells up and releases the water as and when the plants need it. Other water-retentive materials such as small pieces of polystyrene can be added to the compost mix to aid this process.

Although requiring the use of electricity, using automatic water control systems can help regulate the amount of water being used in the garden. Computer systems can be linked to soil moisture sensors and turn on the irrigation systems when the soil becomes dry. Porous pipe irrigation systems are ideal for use on vegetable plots or within flower beds as they are connected to a hosepipe and linked to a water source such as a water butt. A pump may be needed within the butt to ensure a continuous trickle of water. The pipe allows water to ooze out slowly and steadily along its length whenever the water source is turned on. Porous pipes are most effective when buried under the surface of the earth close to plant roots. On vegetable plots or where flexibility is needed, the pipes can be left on the surface and moved around when necessary. The only drawback is that hoeing or digging has to be undertaken with care since it is important not to cut through the pipes. A clear marker is needed to show exactly where the pipes are located within the garden.

Keep watch on weather forecasts – if rain is forecast then don't water. When watering it is important to choose the time of day carefully – early morning or at dusk is far better than during the heat of the day since less water will evaporate. Aim the water into the soil dishes created around vulnerable plants rather than spreading it haphazardly across the ground. Use watering cans wherever possible since this ensures the most accurate watering direction.

Check the plants and soil before watering and only do so if necessary. Depending on the size of the plant, use a trowel or spade to dig a small hole beside it; if the bottom of the hole is damp, then there is no need to water. Take a look at the condition of the plants. If the leaves are starting to droop then they are beginning to suffer from a lack of water.

Only shallow rooting plants, or newly planted ones, require watering every day during dry spells. For most plants, watering thoroughly every few days will encourage the development of deeper roots capable of coping with dry spells much more easily. The Royal Horticultural Society recommends about 24 litres of water per square metre (5.2 gallons every 10 sq ft) approximately every seven to ten days.

Rain water recycling systems

Rain water is a free resource, and is much better for the garden as it is free of chemicals such as chlorine. Storing and recycling rain water for use during dry spells is essential. As far back as 2000 BC, it is known that people living in the Negev Desert (now in modern-day Israel) captured water from the hillside and stored it in cisterns for later use. Community cisterns were common, creating some that held as much as a million gallons of water underground. Throughout history, methods of capturing rain water have always been common, such as the integral systems built into medieval castles including Orford and Carreg Cennen. In the modern world, the concept of rain water harvesting is undergoing a major revival, both for domestic use as well as large-scale building projects such as the Velodrome at the London Olympic Park, which was deliberately designed to include a rain water harvesting system.

Within a garden environment, butts are the most important source of water conservation since their use enables gardeners to store winter rain to reuse during the summer. Just one water butt can collect 160 litres of rain water. Butts are available in all shapes and sizes, free standing or with a flat back that can be fastened directly against a wall. Styles range from traditional circular barrels to tanks and beehives. Most butts are green in colour, but it is possible to find terracotta, grey or black versions.

When installing a water butt, it should be raised off the ground using bricks. The height of the stand can be adjusted to suit requirements, but should be sufficiently high enough to allow a watering to be easily filled from the tap at the bottom. Two or more butts can be joined together using connectors. This enables water to flow from one butt to another, thus preventing one from overflowing and wasting rain water.

Above left: *Rain water draining from gutters and collected in a water barrel.*

Above right: *Linked water butt storage ensuring no rain water is wasted.*

Ideally a water butt should be attached to every downpipe around the house, as well as an overflow butt in each location. Installing guttering around the edges of sheds and greenhouses allows water to be collected easily from those sites. Placing lids over the top of the water butts will reduce evaporation during the summer, as well as being a safety measure to prevent access by young children, birds and animals.

Keeping gutters clean is equally important. Dirt and leaves can cause blockages, reducing the amount of water being directed into a butt. Using rain water filter collectors will allow any debris to be removed from the butt, as well as reducing algae growth. The mesh cartridges can easily be removed for cleaning when necessary.

Vegetation growing in a blocked gutter.

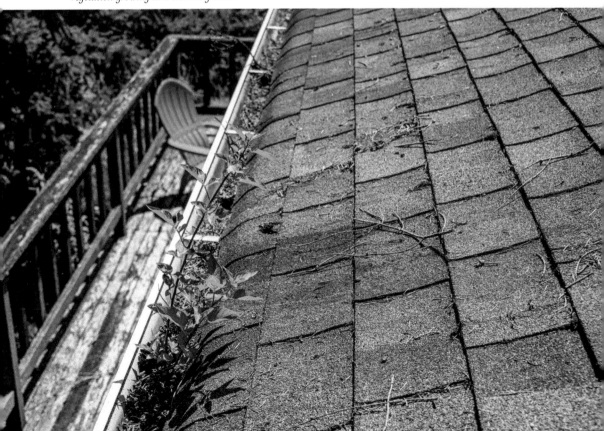

Apart from using a watering can to drain off water for use in the garden when needed, it is possible to set up automatic watering systems be linked to a butt using soaker hoses (also known as porous pipes) to provide regular watering of selected plants. When turned on, water passes through the hose, trickling out at regular intervals through the holes in the pipe. Alternatively, there are submersible pumps that can be placed in the water butt and linked to hosepipes for use in watering hanging baskets or larger areas of the garden.

On a much larger basis, underground storage tanks (cisterns) are increasingly being used. Installing such underground tanks to provide water for use within the garden is not a new phenomenon. At Ickworth House, hidden in the basement walkway at the back of the house, beside the wall of the massive dome, are some metal covers and a nearby wall tap, which would certainly go unnoticed by visitors to the gardens. In fact, these are the only visible signs of the presence of a massive underground tank system used to hold 1,000 litres of rain water that have drained from the house gutters. The tanks are historic, having been installed many years ago.

Head gardener Jack Linfield commented, 'Watering the garden has involved one guy for ten hours a week doing nothing but connect a hose to the tap and move it around. It has to be done in the heat of the day, because that is when someone is available to do the job.'

Recognising that this was not a very effective, nor sustainable method, the National Trust decided to introduce a computerised automatic underground porous pipe system that flows around the entire garden and links directly to the storage tanks. Dividing the garden into four zones, the gardeners are able to programme the watering on a seasonal basis, and can override the system to allow watering only when necessary.

Underground cisterns are ideal for installation when creating a new garden, or replacing driveways. Once installed, little maintenance is required. Such fully enclosed underground tanks are fed by pipes linked to drainpipes that bring water from the roofs of sheds, garages or houses. Placed several feet below the earth, the layers of soil help keep water temperatures, low thus reducing evaporation. Such underwater tanks are completely dark, and water remains at a constant temperature of around 4°C all year. When required, water is pumped from the tanks through pipes to taps, to which householders can link porous pipe systems and hosepipes or use to fill a watering can. These tanks can make a significant impact on water usage. The UK Rainwater Harvesting Association says that such systems can replace up to 50% of household water and 80% used in commercial buildings.

Such tanks are frequently installed when new driveways or paths are laid. The use of permeable membranes and permeable block paving allows rain to pass through the surface, either directly entering the earth or being directed into an underground tank. Permeable block paving can prove extremely attractive as it is usually laid in interlocking patterns.

RECYCLING IN THE GARDEN

Underground tanks being used to store rain water for future use.

Using such permeable surfaces reduces the risk of flooding around the outside of the house, or on nearby paths and pavements. Hebden X Grid is a typical product used for this purpose. Made by British Recycled Plastic, it can be used on gravel pathways and boggy gateways, as well as being part of a sustainable urban drainage system. It helps reduce surface pooling as it creates a grid through which the water can drain.

If you have a small garden, but a large roof or flat area, it may be worth considering installing a rain water harvesting system incorporating a special filter. This filter separates debris such as leaves from rain water. The water then passes through a covered grille section on the drainpipe and into a collection tank. Leaves and dirt are left to fall into the drain. Such water can then be used in a washing machine or to flush toilets. Water saved in this way is not suitable for drinking unless further treatment is undertaken, which can be expensive. Payback periods for installing a rain water harvesting system can take time, however it has been suggested that storing rain water for toilet flushing, use in washing machines and garden watering can reduce water charges by a third.

Reinforced grass, often linked to underground cisterns, is a popular water-saving measure and is particularly suitable in environmentally sensitive locations. It can also look good on front gardens where limited space requires vehicles to parked on what would otherwise have been lawns and flower beds. It can be created in a variety of ways:

Reinforced grass growing through a special mesh.

Block paving can be laid in a pattern involving pockets of soil, which can be seeded with grass or low growing sturdy herbs such as thyme.

Concrete cellular paving can be obtained as pre-formed blocks. The holes in the paving can be filled with aggregate or grass.

Plastic cellular paving is made from recycled plastic laid in an interlocking design. A special base may need to be created or in some cases the paving can be laid directly onto the turf and pegged in place. The cells are filled with soil and grass seed.

Plastic mesh can be used on firm, well-drained ground. It is laid over an area of existing turf or grass. This allows frequent use without damaging grass roots.

In each case, the grass can be mowed as required during the growing season and creates a very hard-wearing, green environment.

Like reinforced grass, gravel can be laid on top of soil covering an underground rain water tank. Wheelchair and pushchair users can find it hard to move on gravel, which is why resin-based systems that bond the gravel together to create a porous, yet solid and smooth surface are popular options.

Driveways, patios, terraces and paths made from recycled materials that can be linked to sustainable water systems are increasingly common. One such material is Oltco's Recycle Bound, a unique resin-bound solution made using waste plastic obtained

Oltco's recycled plastic paths at the Eden Project. (Oltco Limited)

from plastic recycling points. Recycle Bound surfaces are made using a combination of plastic drink bottles, plastic food packaging and plastic straws. Hard wearing, non-slip, easy to clean and extremely durable, every square metre of Recycle Bound contains the equivalent of 3,000 plastic straws, ensuring that a standard 50 sq m drive involves the recycling of 150,000 plastic straws. As the material is totally porous, water drains away easily from the surface and can be linked into sustainable urban drainage systems. Recycle Bound plastic resin pathways have been installed within the biomes at the Eden Project in Cornwall, creating an extensive network of paths that are low maintenance, sturdy and environmentally friendly as they use the equivalent of 255,000 straws.

More unusual, decorative but equally effective water-saving systems are Japanese devices known as Kusan Doi. Designed to celebrate the rain, they create a temporary water feature out of every rainfall. Traditionally, Kusan Doi were formed from a series of small metal cups with holes in the bottom. Hung vertically in a group along a chain, gravity ensured that the water fell from one level to another. Similar methods can be created using a variety of items such as groups of keys, lengths of chains, long lengths of wire spirals and even mugs, ceramic tubes and small tins. In his book *Garden*

Eco-chic, Matthew Levesque describes creating an unusual Kusan Doi using scrap metal. He came across more than a hundred 4in copper squares, each containing a 3in circle in the middle. Upon enquiring further, Levesque discovered that these squares were leftover materials from an artist who had required a selection of copper circles. The remainder was simply scrap metal. Taking the squares home, Levesque turned them into a decorative rain chain joined by small brass rings.

Another version of the rain chain concept was used to combine a decorative effect with water storage. A chain carried rain water from the roof to a drainage hole in the ground, linked to an underground tank. During periods of heavy rainfall, the water cascaded down the chain, creating the impression of a small waterfall. When needed in the summer for use on the garden, the rain water could be pumped to the surface from the underground tank.

Above left: *A traditional metal cup rain chain.*

Above right: *A chain taking roof water into underground storage areas.*

Grey water

It is not just rain water that can be recycled for use in the garden. Grey water that results from household tasks such as cooking or washing may be suitable for use on plants of all kinds, including the vegetable patch, flower beds and around trees. It can provide a significant quantity of water, especially during periods of drought or long dry spells as this can be a difficult time for recently planted trees, shrubs and vegetables. Reusing grey water for any other purpose than on the garden is extremely difficult. Grey water will contain germs, bacteria and cleansing agents. The water cannot be stored for long periods as it will soon start to smell and turn rancid. Although it is possible to obtain small-scale water treatment systems, the general recommendation is that the amount of energy and resources required to operate these can far outweigh any benefit.

There are various types of grey water available for recycling:

From cooking, preparing food and cleaning
From washing and rinsing clothes
From use in personal hygiene such as baths, showers and hand washing
From washing the car (rinsing water only)

When recycling this water, and using it in other ways, it is important to bear in mind the original purpose for which the water was used.

Water used to wash vegetables could be recycled.

Water in the kitchen requires extra care when recycling. Hot water needs time to cool before reusing as the heat could damage plants. Water used for steaming, or boiling eggs and vegetables, will contain additional nutrients that are extremely valuable to growing plants. This water can be used anywhere in the garden as soon as it is cool to the touch. Always bear in mind that the clearer the water, the more ideal it is for vegetables and delicate plants.

Dishes should be washed by hand rather than using a dishwasher. This method uses less water, and by using washing-up bowls it is easier to capture it. Use two bowls, one for washing and one for rinsing. Washing-up water resulting from cleaning plates and other kitchen items should be poured away rather than used on the garden. Such water is generally very dirty, containing grease, food particles, oil and chemicals. The rinsing water is another matter – that is much cleaner and can be used to water plants.

Always check the water quality before reusing on the garden. Water that contains salt or chemicals such as phosphorus should not be used. Such water will rob the soil of its fertility rather than help it. Consequently, it is essential to check the ingredients of soaps and detergents to ensure that they do not contain unsuitable chemicals. Choose soaps and detergents that are as mild and biodegradable as possible.

Grey water that was previously used for washing clothes or personal hygiene will contain small amounts of soap and detergents but generally these are sufficiently diluted so as not to cause any problems. It is important to remember that such water can also contain pathogens and bacteria. If someone in the household has been ill, their bathwater or washing water should not be used on the garden until they recover from their illness.

Water from baths, showers, washing machines and sinks can be removed using buckets and bowls. This can be time consuming, and hard work – especially if you are carrying buckets of water down the stairs and into the garden. Alternatively, simple diverter valve kits can be installed to allow grey water to be diverted from downpipes into hosepipes or porous pipe irrigation systems.

Storing grey water is not recommended due to the potential pollutants and bacteria that are present. It will quickly start to smell, so as soon as it is cool it should be used immediately on the garden.

When using grey water within the garden, always use it with care. Try not to splash any on leafy crops such as lettuce. Wherever possible, aim to pour the water close to the roots. All crops that have been grown using grey water should be washed thoroughly before use.

Large landscaped gardens such as those surrounding offices, schools and community areas can benefit from the installation of reed bed systems. This involves creating a wetland area where filters such as aquatic plants and water-borne life clean pollutants through natural processes. These systems can take up to a year or more to

RECYCLING IN THE GARDEN

A reed beds drainage system at Beeview Farm, Pembrokeshire. (Matthew Watkinson)

establish properly, and require a large amount of space, although it is not unknown for smaller-scale systems to be created with the aid of some imagination. Living in a field with no permanent infrastructure allowed, Matthew Watkinson of Beeview Farm created an extremely effective reed bed system using a group of linked tanks that were capable of dealing with all the family's waste liquid from showers and sinks.

One of the more unusual aspects of grey water recycling is linked to air conditioners that create water as a by-product. Known as condensate, it is most common in hot, humid conditions. Evidence so far indicates that this condensate can be used carefully within gardens. Microsoft uses condensate to irrigate its campus landscaping as well as cooling buildings at its offices in Twycross, Leicestershire; Hyderabad, India; and Herzliya, Israel. Likewise in Austin, Texas, Austin Water promotes condensate reuse on a large scale involving a residential skyscraper that creates 12,800 gallons of condensate, which is then used to irrigate a tenth-floor green space.

Energy Recycling

Gardens offer tremendous potential for energy efficiency – an important factor in today's world.

Natural energy resources such as oil and coal, upon which we have depended for so long, are being used up at an extent that is unsustainable. Such resources are finite, and cannot be replaced since these fuels were created thousands of years ago. These are resources that are not just used to provide heating, lighting and transport but are used in almost every aspect of our lives, including within the garden. Petrol-driven lawnmowers are an obvious example of fuel use but there are many others that may not be quite so apparent. Oil, for example, is used in the manufacture of almost everything used within the garden, such as plastic bags, lawnmower casings, sticky tape, paints, varnish, tools, insulation, electric sockets, watering equipment and plastic garden furniture.

A traditional manual push mower ready for use.

The implications for gardeners seeking to become more energy efficient are significant. Choices have to be made, and it is important to look closely at the ingredients when buying many of the items used automatically around the garden. Looking for the most sustainable options, recycling and reusing materials can make a dramatic difference. Opting for recycled plastic garden furniture, or seeking out pre-loved furniture at car boot sales or garage sales, will affect the demand for the production of new and single-use plastic, thus reducing the use of oil. This is just one example, but the more people who make these choices will mean that the impact will be much greater worldwide. One gesture made by many people ultimately has a cumulative effect.

Quite apart from the energy used in the production of garden tools and other equipment, every task undertaken within a garden requires the use of energy, whether it is simply digging and hoeing, mowing lawns or the construction of the tools and equipment required for garden-related tasks to be completed. Gardening requires the use of considerable personal energy, whether it is walking behind a mower, pushing it or doing a workout in the form of a session of digging, weeding and pruning. The impact of such energy output can be magnified by equipment choices such as using recycling mowers designed to cut grass and mulch the ground at the same time.

On a simple practical basis, rechargeable batteries are an extremely energy-efficient means of powering garden machinery such as lawnmowers and hedge trimmers. When no longer needed, the batteries can in turn be recycled. Gardeners with a small area to cultivate could consider by far the most energy efficient method of all – human power. Manual lawnmowers, hedge trimmers and clippers can be very energy efficient, as well as helping gardeners stay fit!

Fuel costs are rising steadily, and it is becoming much more expensive to use petrol-powered mowers or use mains electricity for outdoor lighting and other electrical systems. Given this situation, gardeners are increasingly seeking alternative energy sources.

Solar power

Every day, more solar energy falls on the earth than is used by people worldwide. It provides a continual source of energy. In 1839, scientist Edmond Becquerel experimented with electrolytic cells, resulting in the discovery of the photovoltaic effect. Following further research, the first solar cells were created in 1883, and ten years later a patent for a commercial solar water heater was issued. By the late twentieth century, solar panels and solar products were becoming widespread, and technological innovations have made solar energy much more effective within northern climates.

Solar power is now used to provide extensive illumination around the garden, including functional lighting, passive infrared security lights, party lights, Christmas lights, decorative lighting and fountains with integral solar pumps or automatic watering systems. Individual cells contained within each product store up solar energy, releasing it when required using manual or automatic switching systems. Even in the UK, solar lights can be very effective all year round as there is sufficient daily sunlight to recharge batteries to between 50% and 100% capacity. The brighter the sunlight and the longer the sun shines, the higher the charge received by the battery. This allows it to operate for longer periods. Most solar lights will operate for around eight hours given reasonable sunlight. On/off switches allow householders to turn

Solar lights within a garden border surrounded by tagates.

lights off to store up energy for use when required. Wipe off dirt and dust periodically as this will dim the amount of light being given out, as well as reducing the amount of energy passing to the integral battery.

Careful placing of greenhouses, sheds and cold frames can make a tremendous difference to the amount of energy being used within a garden. Use a compass to identify which direction is south, since this will receive the most sunlight. Maximising energy conservation by angling cold frames and greenhouses to use that natural energy will ensure that plants benefit from maximum heat during winter, spring and autumn.

A greenhouse can be very costly to run, especially if you are planning to heat it during the winter. Such costs can be decreased through careful location and recycling methods. Wherever possible, greenhouses should be placed in a sunny, unshaded location in order to benefit from maximum solar energy. The glass should be cleaned at regular intervals, especially in the winter, since dirty glass reduces heat and light levels. A lean-to greenhouse located against the side of a house or a garden wall will benefit from interior heat transmitted through the brickwork.

For generations, gardeners have used a very simple technique to provide free heat for greenhouses and walled areas. Painting an interior wall white traps heat in the wall, encouraging plants to grow and ripen for longer periods and resulting in extremely productive cropping periods.

Above: *A greenhouse placed to maximise energy gain from an adjacent building.*

Below: *Painting walls white encourages heat retention, which benefits climbing plants. (Angela Youngman)*

A geothermal heat system showing how it affects the garden.

Other sources of energy

A far less well-known source of energy recycling is ground source heat (also known as geothermal heating). This method of heating uses all the natural heat contained within the ground. Within the UK, just a few metres under the soil the ground maintains a level temperature of 11 to 12°C. This heat can be transferred from the ground into a building to heat radiators and provide hot water by using a system of coils and pipes, which are placed either vertically or horizontally into the ground. For every unit of electricity used to operate such a ground-source energy system, it has been estimated that three to four units of heat can be produced.

Installing ground-source heating of this kind is best undertaken when creating a new garden as it does involve considerable upheaval as well as restrictions on the type of planting that can be undertaken. Deep-rooted trees and bushes cannot be planted near the system since the roots would ultimately affect the system as it can only sustain low-level planting such as lawns, heathers or annual flowers.

Micro-turbines to provide wind energy can be obtained but these do require professional installation, as well as a pre-site survey to ascertain that there is a high enough wind speed to sustain the turbine. Planning permission is required. Such small turbines are often used on allotments to power batteries and electrical products.

Inspiration from the past

Seeking inspiration from historic gardening practices, or traditional ones, can provide many useful ideas. In Victorian times, gardeners were extremely resourceful, intent on

Seedlings being grown on top of a compost hot bed.

using every possible source of energy and space. During the 1980s and 1990s, Harry Dodson, a retired head gardener, took part in BBC TV series recreating traditional kitchen garden techniques within an old walled garden at Chilton Lodge in Berkshire and discovered historic ideas could be very productive. During the series *The Victorian Kitchen Garden*, he experimented with making a Victorian hot bed using horse dung. Hot beds were used as a way of growing crops under cold frames. Having obtained sufficient quantities of horse manure, it was piled up and turned up to five times throughout a fourteen-day period. This kept the heap from becoming too hot for cultivation. Leaves were mixed with the manure to further moderate the heat, and it was then compacted down to a depth of just under a metre (3ft). A cold frame was placed on top. It was then left for five days to allow the rank steam to escape, and then a loamy soil layer was placed on top of the manure. The soil quickly became warm enough for planting, and a crop of short-rooted carrots was grown successfully. Writing in the book about the series, *The Victorian Kitchen Garden*, Jennifer Davies stated, 'Provided that they were kept at 70°F (21°C) by lining with fresh dung and making sure they were covered over in cold weather, hot beds could be kept in production for a considerable time.'

It is not just purpose-built hot beds that could be used for this type of cultivation. Visits to allotments have revealed many modern gardeners taking advantage of the heat generated by compost heaps to grow pumpkins, squashes and even tomatoes. By placing a layer of soil on top of the decomposing heap, and planting into that soil, allotment gardeners have been able to successfully grow numerous young plants resulting in harvests of large, succulent fruits. When harvesting is complete, the compost heap can be added to the vegetable beds, while the dead plants become part of the next heap to be created. Potatoes are another crop that are frequently grown on top of compost heaps.

Placing cloches on top of compost heaps in early spring will encourage germination. The cloches act as mini greenhouses, enabling the young plants to thrive due to the additional heat provided by the presence of the compost heap underneath their roots. This method has been used in sealed compost bins, as long as the seeds being germinated require darkness and high temperatures to succeed. Once germination has taken place, the lid does have to be removed to allow the plants to continue growing.

Composting

Compost is by far the most common form of energy recycling as it turns unwanted materials into free compost and soil conditioners. It is the most environmentally friendly method of dealing with kitchen and garden waste and is suitable for all gardens. Varying types of composting systems are available that should meet all requirements. For households that are unable to create their own compost due to a lack of space, councils offer green waste collections by providing a special bin at an extra charge. This is emptied by waste disposal vehicles on a regular basis.

Compost heaps

There are a variety of compost bins available. These include home-made constructions using pallets or wooden boxes topped with old carpet to keep the heap warm. There are round tumbler bins and plastic bins of all shapes and sizes. Some bins are fully enclosed, while others just rest on the earth. Everyone has a different view as to what type of bin they prefer – much depends on the availability of space, access, physical strength, time and effort, as well as personal preferences. When choosing your compost bin, ensure that it will keep out rain, retain some warmth, drain away surplus liquid and allow air to circulate. Whatever the type of bin chosen, the basic techniques of creating a compost heap is always the same. You have to layer the material, turn it regularly, keep it moist and remove the contents when they have decomposed into a friable soil.

The location of a compost bin is a key factor in ensuring its success. The Royal Horticultural Society points out that 'it is important that the site is not subjected to extremes of temperature and moisture, as the micro-organisms (bacteria and fungi) that convert the waste to compost work best in constant conditions. Position the bin in light shade or shade; it is often more convenient to use a shady area of the garden.'

Ideally gardeners should have two compost bins available for use. When one bin is full, it can be left to decompose while the other is being filled.

Creating viable compost takes time and effort, as it is important to achieve the correct combination of waste materials.

Kitchen waste being added to a compost heap.

There are two types of compostable waste material: green and brown waste. These should be added to the pile in alternate layers.

Green waste comprises:

Uncooked vegetable peelings, salad and fruit
Fresh garden waste such as grass clippings, hedge trimmings, leaves, pruned material and decayed vegetable remains
Dead indoor plants or flowers
Flower stems
Fresh farmyard or stable manure
Bedding and container plants after they have died at the end of a season

Brown waste comprises:

Twiggy material
Pruned wood clippings
Shredded paper, especially newspaper
Egg boxes, cardboard tubes
All types of cardboard ripped into small pieces, having first removed any staples or Sellotape
Teabags, loose tea or coffee grounds
Egg shells
Wood ash
Straw and hay
Hair

There are materials that should not be added to a compost heap, such as:

Meat and bones
Fish
Dairy products
Cat and dog litter and waste
Large logs or pieces of wood
Coal ash
Diseased plants
Perennial weeds and weeds in seed
Glossy paper
Soot
Cooked food

Human and animal hair is possibly one of the more unusual ingredients of compost bins since it is a valuable source of nitrogen. Jennifer Davies, author of *The Wartime Kitchen Garden*, notes that 15½lb of hair produced about 2¼lb of nitrogen and recalled the experience of Muriel Smith, a young apprentice hairdresser in Cirencester. Muriel stated that after she left school,

> I was apprenticed to ladies hairdressing ... There was a patch of garden at the rear of the salon, and here we dug deep trenches lined with newspapers, plus thick layers of hair cuttings. We saturated this well, filled in with earth and planted runner beans. They did remarkably well ... the yield kept us busy, slicing and salting down whenever there was a moment to spare between clients.

It should be said that over time, views on suitable compost materials, especially household waste have changed. Writing in the 'Victorian Kitchen Garden', Jennifer Davies notes that Mr F.C. King, head gardener of Levens Hall in Cumbria, placed everything into his wooden compost bins, including garden waste, household waste, 'even cinders, ash and clinkers'. During the Second World War, gardeners were encouraged to ask chimney sweeps for soot. When combined with wood ash, it produced a good fertiliser.

Having added green and brown waste to a compost heap, it is important to turn it periodically. This involves the considerable physical activity of digging out the

Turning a compost heap to encourage decomposition.

compost by moving the outside waste into the centre and vice versa. Many gardeners simply turn all the compost out of the bin and replace it as it is mixed together. Enclosed rotary bins have handles that can be turned, rotating the bin and thus circulating and mixing the material.

It is important to remember that composting takes time and requires patience. A compost heap does not turn into useable compost overnight. The process of turning a heap provides an opportunity to assess its progress and deal with any problems as they occur.

What are the most common problems?

Wet heaps

Sometimes a compost heap becomes too wet and turns into a soggy mess. This is usually due to the fact that too many grass clippings and elements of green waste have been added. The answer is to add plenty of dry twiggy material and other brown waste, before mixing thoroughly. Future problems with wet heaps can be avoided by mixing grass clippings with some dry brown waste before adding to the compost heap. A good way is to collect up autumn leaves, small twigs or cardboard and store in bags nearby ready for use when needed. Add some to the grass mix before placing in the compost bin.

Smelly heaps

A compost heap is composed of decomposing materials. It can get extremely smelly, especially if the heap has become too wet. Rotting grass clippings create a strong smell of ammonia. To solve the problem, mix the contents with lots of bulky dry brown waste such as leaves, straw or twigs. Mix well. Then cover the waste with partly composted mulch from the bottom of the pile or a little garden soil.

Dryness

If a heap becomes too dry and cold, it stops decomposing. At this point, there is only one solution. You have to empty all the material out of the container and begin again. Refill the bin, making sure that all the outside material is pushed to the centre of the compost heap. Add some water and some nitrogen-rich material such as grass clippings, manure or kitchen scraps. Mix the heap thoroughly. Place thick layers of cardboard around the edges of the compost heap – this will help keep warmth within the heap and encourage decomposition.

Slow decomposition

The presence of large quantities of long branches or thick stems is the usual cause. You need to cut the material into small pieces or place them in a shredder before adding

to the compost heap. Make sure that all fibrous stems and leaves such as those from sweetcorn plants are cut up thoroughly before mixing into the heap.

Pests

Ant hills are frequently found in compost heaps. They are a perennial problem and are virtually impossible to avoid. All you can do is open them up and leave the heap uncovered for a while to allow birds to eat the ants and their eggs. Pouring water on an ant hill will kill the ants but bear in mind that large quantities will be needed, especially if it is a large heap. Check the weather forecast to see if any storms or heavy rain is forecast. Opening the heap and removing the lid will allow in the rain as well as give access to birds. However, this will also make the heap soggy, which means additional dry material will have to be added quickly in order to restart the heap and increase the heat levels.

Mice and rats have been known to make their homes in a compost heap. Turning the heap regularly will act as a deterrent. Avoid adding any cooked food to the decomposing heap, especially meat and bones. Uncooked food such as vegetable peelings should be placed in the centre of the heap and kept covered.

Flies are an inevitable problem during summer. They appear in large numbers whenever the compost has become too rich in nitrogen or there are significant quantities of food scraps present. Dealing with the problem requires stirring the compost regularly, and covering any food scraps with a thick layer of green waste or soil.

Alternative forms of composting

Different types of composting depend on someone's personal requirements and the amount of space available.

Hot composting

Hot composting can be undertaken through the creation of a standalone pile, or within a plastic hot composting system such as the Green Johanna or Green Cone. Heat is used to accelerate decomposition, with interior temperatures capable reaching up to 49–77°C (120–170°F) within a few days. The temperature of the pile has to be monitored carefully, since if it becomes too hot it will kill the beneficial micro-organisms needed to develop the compost. The Permaculture Association recommends using a compost thermometer or a meat thermometer attached to the end of a stick.

When creating a hot pile, use equal parts of green and brown waste, ideally shredded to a small size such as a mix of fresh grass clippings and dried shredded

Thermometer showing the level of heat that can be generated within a compost pile.

leaves. Mix together and add a shovelful of compost or soil to encourage the growth of micro-organisms. Water should be sprinkled on at regular intervals as the pile grows, creating what the Permaculture Association describes as 'the consistency of a wrung-out sponge'. Continue adding to the mixture until it is around a cubic metre. At this point, the mixture should be left and the temperature monitored on a daily basis. Within five days, the interior temperature should rise to 49–77 degrees Centigrade (120–170 degrees F). The exact temperature will depend on the level of moisture, the size of the pile and type of organic material. When the temperature shows signs of cooling, the mixture should be turned to bring material from the edge to the centre of the pile and allow the introduction of oxygen, which will reheat the pile. This will also ensure that all the material decomposes evenly. In general, the pile will need to be turned every four to five days. When the mixture is a dark, crumbly compost, it can be allowed to rest for a couple of weeks before being used.

A plastic container such as the Green Cone composting system acts as a food waste digester, using solar heat to reduce leftover food into a nutrient-rich liquid, which feeds the surrounding soil. It has to be placed in a sunny, well-drained location in order to be effective and is particularly useful when sited within a vegetable plot or allotment. The Green Cone is not suitable for placing on patios or a concrete base as it must be dug into the earth.

The Green Cone comprises a rigid green plastic container with an integral thick plastic floor, which has to be underground. Cooked food waste, meat and bones are deposited into the interior of the Green Cone. Care has to be taken to ensure that it is not filled above the buried part of the bin and that only food waste is placed inside. It is not

RECYCLING IN THE GARDEN

suitable for any other form of green or brown waste. The warmth of the sun and the earth raises the temperature of the plastic container to a very high level, thus enabling the food waste to decompose naturally. No stirring or mixing is required. Pests such as rodents cannot get into the Cone to eat the waste material. The excessive heat reduces all the waste produce into a liquid, which seeps through the plastic net and into the soil below, thus increasing soil fertility. Special accelerator powders can be obtained to add to the cooked waste in order to encourage the development of a healthy bacterial population.

Digesters

Wormeries and Bokashi systems are particularly suitable for dealing with smaller quantities of kitchen waste.

Wormeries are a very efficient form of composting kitchen waste. Each wormery contains at least two sections: a sump at the bottom from which liquids can be drawn regularly, and an upper area containing moist bedding material that creates a humid area where the worms can begin to burrow and digest their food. A thin layer of kitchen waste is added to this upper layer. After about a week, the worms will have settled in and munched their way through the materials. At this point more kitchen waste can be added. When the upper section is full, it is covered by another basket, into which waste materials are deposited. Worms eat the material in the lower basket, and as they complete the task they steadily move upwards and into the next basket. When all the worms have moved to the top basket, and the contents of the lower basket have become compost, it is emptied and used to provide a new top basket ready for refilling. Special types of worms are used in wormeries. These are known by various names such as brandling, red or tiger worms and are smaller and darker red in colour than the common earthworm.

In order to be an effective method of composting, the wormery must be placed in a sheltered area or in a shed so that it maintains a constant temperature. The worms are most active in warm, moist conditions, with perfect operating temperatures being 18–25C (64–77°F). Their activity will slow down at lower or higher temperatures.

Care has to be taken to ensure that the material does not become waterlogged, especially during periods of heavy rain.

Wormeries use kitchen waste, tea bags, eggshells, coffee grounds and small quantities of bread. Limited amounts of paper and cardboard, together with small amounts of soft green waste such as leaves or annual weeds, can be added.

Wormeries are not suitable for composting woody material, fibrous leaves such as sweetcorn and dairy products, meat, fish and bones.

Some problems can occur when using wormeries. If too much waste is added at any one time, the worms may not be able to cope. As a result, the waste will start to decompose, attracting vermin and flies. Dealing with this is simple – remove the excess waste and wait until the worms begin digesting the top food layer before adding more material.

Using a Bokashi compost system to convert food scraps.

Unpleasant odours can result if the wormery becomes too wet. Excess liquid should be drained away and shredded paper or card added to the compost mix to absorb excess moisture and increase air circulation. Also, check that the drainage holes are clear and that the worms are still active. If the problem continues, the compost mix may have become too acidic and a small dressing of calcified seaweed or calcium carbonate will be required.

If the waste material are no longer being eaten the reason is simple: the worms are no longer active. Check that the holes between the layers are clear, so that they can move between the baskets. Take a close look in the baskets to see if any worms are present. If none can be seen, then you will need to obtain more from the supplier. Digging up some earthworms from the garden will not work – you do need a specific type of worm to make a wormery effective.

Bokashi composting systems are designed for indoor use and aim to decompose small quantities of cooked food, fish and dairy products. Each composter comprises a sealed unit in which each new layer of food is covered by a layer of Bokashi yeast. Liquid residues have to be drawn off via a tap at regular intervals. This liquid is extremely nutritious and useful as a fertiliser, as long as it is diluted before use. When the container is full, it has to be left until the food decomposes into compost. Since this can take some time, most households using this method have two or three Bokashi containers so as to ensure that one is always ready to be filled.

Green Roofs and Green Walls

Covering walls and roofs with greenery has become an increasingly popular form of energy and water recycling. This practice is nothing new since such techniques have been in use for centuries. Built around 500 BC in what is now Iraq, the Hanging Gardens of Babylon were the most famous example of a building covered in vegetation. The gardens were designed by King Nebuchadnezzar II to provide his wife with an environment that would replace the greenery of her homeland. The resultant gardens were grown over stone pillars and roofs, which had first been waterproofed with layers of reed and tar. So unusual were the gardens that they became renowned as one of the Seven Wonders of the Ancient World.

Green roofs on houses beside a fjord in Norway.

Archaeologists have discovered roof gardens within the ruins of Roman Herculaneum, buried under the lava flow from Mount Vesuvius in AD 79. The Vikings used sod and grass to cover the roofs of their houses, as did eighteenth- and nineteenth-century settlers in North America who constructed 'sod houses' on the western plains. Faced with a lack of trees with which to build cabins, settlers cut slabs of grass sod to use as bricks. These slabs were placed root-side upwards so that they would grow into the next slab, thus developing a more solid foundation. Similar sods were used to create roofs.

In Scandinavia, green roofs have been used to provide wild flower meadows and grow alpines. Traditional log cabins frequently incorporated a green roof designed to provide winter insulation, while keeping the cabin interior cool during the summer. Construction was simple: plants and soil were kept in place using wooden boards across the roof. Alternatively, the roofs were covered with a layer of turf. These green roofs were waterproofed by placing layers of birch bark above a sealed plank surface. A layer of birch twigs was placed on top of the bark in order to allow water to drain away. Turf and soil was placed on top and then held in place with boards.

Flat green roofs were being constructed on houses in Germany during the early 1900s, and by the mid-1930s a recognisably modern-style green roof was constructed on the Rockefeller Centre, New York. The energy crisis of the 1970s encouraged Germany to begin exploring the energy conservation potential provided by green roofs, and within thirty years there were an estimated 13 million sq m of roofs

The dramatic Hundertwasser house in Vienna with its green roofs and walls.

A green roof on top of a shed.

covered with greenery throughout the country. It was a trend that crossed national boundaries and attracted worldwide attention.

The dramatic buildings created by Austrian architect Friedensreich Hundertwasser frequently involved trees and bushes growing out of walls, green roofs and vertical green installations as a way of ensuring his projects harmonised with nature, diminishing the visual pollution of the environment. An example of this is the Hundertwasser haus in Vienna, which contains fifty-two apartments. Trees cover the roofs and others grow inside apartments, resulting in tree branches sticking out of the windows.

Since the turn of the century, green roofs have become an increasingly popular form of energy recycling worldwide and have been used to cover motorways, schools and shopping areas, as well as residential properties. Within the UK, green roofs have been used on numerous urban buildings, such as the 1,500 sq metre Royal Bank of Scotland building in Edinburgh, 270 sq m of wild flower turf on the Interpretation Centre close to Cardiff Castle, and West Ham Bus Garage in London. In Norwich, the Castle Mall shopping centre is constructed in a sensitive location beside a historic medieval castle. As a result, much of the centre is underground and covered with an intensive green roof that acts as a public park complete with flower bed, shrubs and lawns. Not far away at the Sainsbury Centre, University of East Anglia, a green roof covers a large proportion of the floor space of the art gallery. In Southwold, Suffolk, the 0.6 acre sedum roof on top of the Adnams Brewery Distribution centre acts as a massive rain

Flowers and grass growing naturally on a roof.

water catchment combined with a sustainable urban drainage system, allowing the company to harvest rain water from the roof.

Green roofs are equally popular on smaller domestic projects such as garden sheds, garages, outbuildings, pergolas and even dog kennels and beehives. It is easy to see why – this is a viable way of using space that would otherwise be wasted, while recycling heat and water. Green roofs are environmentally friendly, providing habitats for wildlife, especially birds, thus increasing biodiversity within an area. Other advantages include protecting the roof from damage, improving sightlines and vistas within gardens, and adding visual interest to buildings and structures. A green roof helps absorb CO_2 from the atmosphere and helps filter dust and pollution, while providing insulation against heat and cold.

The basic definition of a green roof is that it is covered with some form of vegetation. Current gardening practice involves three types of green roof styles:

Naturally occurring green roofs occur over time. Nature will automatically take advantage of bare soil to allow plants and grass to grow. Such roofs tend to be mainly grass and wild flowers.

Intensive green roofs reflect conventional gardening styles. They comprise a typical garden complete with trees, paths, shrubs, seating areas and even occasionally water features but are located high above the ground. These gardens do require much more maintenance and the buildings involved have to be capable of bearing a considerable weight. A flat roof is normally required for such gardens.

RECYCLING IN THE GARDEN

An intensive green roof located on houses in Milan.

Extensive green roofs have become the most widespread of all roof garden styles. These generally focus on the use of turf, wild flowers, sedums and low-growing herbs such as thyme and mint. These gardens are used widely on all types of buildings from schools to sheds, private houses, garages, bird tables and outbuildings. Such gardens offer considerable advantages in terms of construction and long-term maintenance as they are much lighter in terms of weight, and can be used on both flat or sloping roofs. If using sedums, gardeners have the option of individually planting lots of plants or using sedum matting – rolls of sedum that fit together quickly to cover a wide area.

A sedum roof combined with solar panels for maximum energy conservation.

Whatever their size and style, green roofs offer significant environmental benefits. Research has shown that they help to reduce overall temperature levels, especially in urban areas. The United States Laboratories Public Health Building in Salt Lake City, one of the hottest areas of America, noted that as soon as a green roof was installed on its building, the interior temperature kept to 70°F when the exterior temperature was 110°F.

Green roofs provide useful insulation. The vegetation reflects solar energy away from the roof membrane, reducing the amount of heat leaving a building in winter. This provides fuel cost savings. During the summer, a green roof provides shade from strong sunlight. Noise levels are also reduced. The vegetation helps to insulate the interior of a building by deflecting or absorbing the noise. A green roof with a 12cm (4.5) substrate reduces interior noise levels by 40 decibels.

They also assist with storm water management during bad weather, reducing and slowing down the amount of rain water that flows down into the drainage systems. Approximately 15 to 20% of storm water is retained on the roof for at least two months before draining away. This reduces the risk of flooding from storm water surges during heavy rainfall. Owners of green roofs have commented that the rain does not actually start running off a roof for at least an hour after it has started raining.

This results in an extremely water-efficient method of landscaping, and can be linked to grey water collection systems where necessary. In Norfolk the owners of a mud-brick eco-shed designed the green roof to ensure that all surplus rain water drained into underground storage tanks. A network of underground pipes pumps the water from the tanks to butts elsewhere in the garden.

Every green roof contains specific basic components: the original roof floor, a waterproof membrane, root protection membrane, drainage membrane, growing medium and planting materials. Many green roof owners, particularly of smaller projects, have recycled existing materials using items such as thick layers of builders plastic as a waterproof membrane, or surplus wood to make an edging framework.

Green roofs should not be placed directly on top of asphalt or bitumen layers as these are susceptible to damage from plant roots. A waterproof layer and a woven fabric such as a geotextile cloth must be used on top of the original roofing materials.

The actual vegetation used on a green roof ranges from turf to perennials, herbs, wildflowers, shrubs and vegetables. Sedums are particularly popular since they can cope with low levels of water as well as being light in weight. This is an important feature to consider when planning a green roof, since any type will add extra weight to the existing structure. Some owners of larger green roofs on top of sheds and outbuildings have added to the biodiversity by incorporating some beehives as well.

Whatever the type of building, it is important to ensure that the weight load strength is not exceeded. Quite apart from the actual green roof, there will be additional weight

Above: *A lush green roof on top of a Norfolk eco-shed used to capture surplus rain water. (Angela Youngman)*

Below: *The layers required by a green sedum roof.*

resulting from water retention, which can be significant depending on the season. All buildings including sheds, wood stores, rabbit hutches and dog kennels should be checked carefully to ensure that the structure is able to cope with the extra weight. In many circumstances, the roof may need to be strengthened to cope with the additional load. Retrofits of this kind tend to involve the use of a wooden or metal framework built around the existing roof with posts situated underneath. Examples seen on allotments include roofs where an additional canopy has used surplus heavy-duty timber posts, strong plywood and old chicken wire to hold the sedum planting in position.

In creating the eco-shed roof in Norfolk, the owners needed over 300 buckets of soil to cover the single-storey, 60 sq m building. A series of ladders and ramps were used to move the soil from the ground to roof height. Once complete, the roof contained a layer of soil, which was approximately 7.5 to 10cm (3 to 4in) deep, together with stones holding down the felt at the far ends and layers of gravel between the sedum planting areas. For longer-term access, a recycled ladder previously used to provide access from ships to a dock was installed against the exterior wall.

Costs depend entirely on size, type of plants and the type of project being undertaken. Recycling materials such as wood and plastic can make a difference to the overall price. Always opt for a high-quality form of waterproof membrane as this is essential in avoiding leaks in future.

Long-term access to the roof is critical in order to ensure effective maintenance. Small projects such as covering the top of a rabbit hutch or dog kennel, will be accessible from the ground, and it is easy to undertake weeding and tidying from time to time. Larger projects on top of sheds, pergolas and outbuildings are more challenging both in terms of the initial access to create the green roof as well as the long-term maintenance.

The larger the roof, the more materials that will be needed to construct a green roof and these can be heavy, too heavy to simply carry up a ladder. A forklift truck may be needed. Even covering a small area such as the roof of a summer house or shed will require the involvement of several people to pass equipment, hold matting in place and above all ensure that ladders are stable and secure. Once the roof is in place, access will be needed at regular intervals to undertake weeding, tidying and keep drains clear. If trees overshadow any part of the roof, leaves will need to be brushed away in autumn otherwise they will affect the fertility and planting.

Using a roof as a garden helps to recycle otherwise unused space, turning it into a useful part of the wider garden. During the Second World War, small-scale green roofs formed an important part of the war effort, enabling countless home owners to grow additional crops. These roof gardens covered air raid shelters nationwide. Writing in *The Wartime Kitchen Garden*, Jennifer Davies writes,

The earth was added protection and it served to make the shelter less noticeable. Garden World (10 August 1940) considered that even the disturbed soil might

A Second World War air raid shelter in a Norfolk garden was used to grow strawberries. (Angela Youngman)

draw machine-gun fire and applauded the camouflage effects of two of its readers, Mrs Prendergast of Clapham and Mr Lamb of Forest Gate, both in London. Mrs Prendergast was growing lettuce, beetroot and marrows on her shelter and Mr Lamb had sown grass on the sides of his and planted the top with flowers.

Other anecdotes indicated that mushrooms grew well inside the gloom of the shelters, as did well-manured buckets of rhubarb crowns in early spring. More recently, old air raid shelters within gardens have been turned into storage sheds, and the roofs covered with wildflowers.

This type of roof coverage could be extremely productive. A Second World War re-enactor in Norfolk experimented by growing crops on top of an air raid shelter within her garden. Strawberries proved to be prolific, growing well in these conditions, while inside the shelter, strings of garlic and onions hung from the ceiling lasted well throughout the winter.

Vertical wall gardening

There is a long-established practice of growing plants such as ivy, Virginia creeper, clematis or clinging vines like a climbing hydrangea across walls to provide a layer of insulation. Such plants reduce the external temperature of a building from between 10 and 60°C to 5° and 30°C. In addition, covering walls with greenery provides much-needed oases of colour in urban areas, adds visual interest, helps local biodiversity and reduces noise pollution as the plants absorb traffic noise. In recent years, the use of green or living walls has become increasingly prominent.

Frenchman Patrick Blanc created his first major vertical garden in 1986 at the Cité des Sciences et de l'Industrie in Paris but it was not until 1994 that landscape gardeners and gardening enthusiasts began to seriously take an active interest in this type of gardening. Tropical rainforests and mountain environments in which plants grow using minimal soil levels inspired Blanc to devise a patented plant wall system suitable for use both indoors and outdoors. His system involves combining a metal frame with a stiff, waterproof PVC layer and capillary matting to take water to the roots of the vertically growing plants. Both climbing and non-climbing plants are used in his green walls, and may be installed as seeds or as pre-grown plants. Automatic watering systems, which can be linked to recycled water harvesting methods, provide the plants with water.

Among the many projects Blanc has undertaken is the green façade of the Musée du Quai Branly in Paris, where an 8,600 sq ft green wall covers the exterior. It contains approximately 150,000 plants comprising 150 species that originate in Europe, North America, China, Japan, Chile and South Africa.

Vertical wall gardens have become increasingly popular, especially in the form of large vertical planters that allow plants to spread across a vertical surface. These wall gardens can be either free standing in a planter, or attached to a wall. These are often watered using a hydroponic system, although some people cover a wall using lots of hanging pots, trellis and boxes that incorporate automatic watering systems. Innovative gardeners tend to use all forms of recycled materials to construct methods to hold plants against a wall, such as long lengths of pruned branches, rope, bamboo and wooden pallets as well as trellis.

A vertical garden on the exterior of the Musée du Quai Branly, Paris.

A simple vertical wall garden created from wooden slats and varying pots.

Recycling Materials

Gardeners are natural recyclers and are constantly looking for ways to deal with problems, tasks and new projects. Recycling also adds a touch of individuality to any garden, providing a personal touch rather than the uniformity resulting from using commercially made pots and other landscaping materials. Such creativity often introduces some fun into the garden design, especially when gardeners transform unlikely objects into planters.

The Gressenhall Farm and Workhouse museum in Norfolk contains a vast array of garden equipment and country tools. Among the collection is evidence of enterprising gardeners from the past recycling tools to provide innovative solutions to garden tasks. Among the objects on display within its galleries are two unique handmade tools. One tool was specially designed to deal with caterpillars on cauliflowers. It comprised a large wide spade with a small area cut out at one side and inwards. The idea was that the gardener would push the cut semicircle under the cabbage to hold the stem in place and then shake the cauliflower, causing the caterpillars to fall onto the shovel surface ready to be taken away and destroyed. The second tool comprised a handmade furrow hand fork. The fork was attached to two strips of wood on either side linked to a wheel. A double handle made from two more strips of wood was placed just above where the fork linked to the first strips. The idea was that gardeners would push the fork through the earth to cultivate it, like a mini-plough.

Over the years, landscaping styles have often proved a way of reusing materials. A typical example was the 1970s trend for crazy paving, which proved to be an easy way of reusing combinations of broken paving, slabs and bricks. With the reawakening of interest in recycling and reusing, such styles and practices are coming back into fashion. Even Chelsea Flower Show gardens have been known to incorporate recycled materials. In 2014, landscape designer Anthea Guthrie's 'A Child's Garden in Wales' show garden used only waste products such as bark edge planks to create a shed and coal slack for pathways, while rusted iron bedsteads and old scaffolding were used to make a junk iron fence.

The RHS Chelsea Flower Show organisers now insist that all garden exhibitors must show where display plants and materials are to be reused. Every garden must live on in some way. A large proportion of the show gardens are donated or sold off for charity. Some are transplanted for use in public spaces, for example the 2019

Crazy paving in a country garden.

Duchess of Cambridge's Back to Nature garden was later installed at Hampton Court Palace, while the Unexpected Gardener display was placed at Thrive's charity headquarters in Reading. In addition to such wholesale transplanting of a garden to a new location, the RHS has linked up with other organisations to ensure that anything surplus to requirements is reused.

During the build-up to the show each year, any materials found surplus to requirements have to be placed in an on-site materials swap shop for any exhibitor to use. Broken tools or tools that are no longer required are sent to Tools Shed, the Conservation Foundation's Tools for Schools recycling project run in association with HM Prisons. After the show, volunteers from the London Community Resource network arrive on site and remove any remaining materials, such as timber, compost, woodchip and plants. They then distribute them to community groups across London. Entire community gardens have been created as a result. Hardcore aggregate is all taken away by Powerday, a London-based waste and recycling business that operates on a 100% recovery basis. Much of the material is recycled immediately, while the remainder is used as fuel to generate energy, waste wood is shredded and soil reclaimed and used on land restoration projects. In addition, concrete is crushed for use in building projects, so nothing is wasted.

Recycling has always been at the forefront of allotment gardening. You only have to take a look at an allotment site to see innovative constructions made from all kinds of materials, such as chunks of metal transformed into plant tunnels or used to hold up netting. Home-made barriers of corrugated iron and varied pieces of fencing often divide the edge of one plot from another on allotments. Sheds have traditionally been built from recycled materials – materials that have been salvaged from many different functions. It is not unknown for sheds and greenhouses to be created out of old packing cases or even part of a sailing boat. Even large quantities of video cases have been used to construct small greenhouses, placing them like bricks in a wall. The

A wide range of materials are being recycled on these Suffolk allotments.

RECYCLING IN THE GARDEN

empty space within each case provide additional insulation. The only requirement is that the resultant constructions are waterproof, safe, stable and will not fall down.

Almost anything can be recycled with a little bit of time, effort and imagination, from food containers to bubble wrap, broken china, tools, compact discs, glassware, wheelie bins, fabrics, metal and rotary washing lines.

Among the most frequently used items recycled in the garden are:

Apple corers

Apple corers are used as dibbers, making small holes for tiny bulbs such as snowdrops.

Aluminium

Pre-cooked meals often come in aluminium containers that can be turned into seed trays. Once these are washed clean and dried, holes can be pierced in the base to provide drainage.

Ash

Wood ash is added to the compost heap.

Ashes of any kind, especially from coal fires, can be mixed with salt and sprinkled on weeds within paths and driveways. The resultant mix will kill the weeds.

Cinder ashes have traditionally been used to as a material for paths.

Autumn leaves

Leaves create an extremely nutritious leaf mould. Rake the leaves into a heap, put wire netting around the heap and leave to decompose. Remove any weeds that develop. Alternatively, place the leaves into thick black plastic bags. Fill the bags as full as possible. Tie up the top of each bag. Punch holes into the bottom of each bag and leave in a sheltered place such as a corner of the vegetable plot. A nutritious liquid from the decomposing leaves will trickle out through the holes and enter the soil. The leaves will decompose quickly, creating a valuable soil conditioner for use anywhere in the garden, or it can be mixed with compost from the heap to make a seed-growing medium.

Turning leaves into extremely nutritious compost. (Angela Youngman)

RECYCLING IN THE GARDEN

Baby baths

These can be turned into small ponds or used as a creative planter.

Bags

Compost bags, growbags and builder's bags offer great potential for recycling.

Cut up and place around strawberry plants as a home-made mulch mat. If the bags are slit carefully along the sides, and spread out, they can be used as a weed suppressant. Fasten down with bricks or pieces of wood.

Pierce compost bags with holes and fill with leaves during the autumn. Tie the tops securely and leave for about a year to rot down.

Fill the bags with home-made compost or compost from wormeries. Tape up the end of the bag and cut holes in the sides. This will enable the bag to be used as a growbag.

Compost bags and other similarly large plastic bags are ideal for storing firewood and kindling gathered up around the garden.

Among the school activities suggested by the Royal Horticultural Society is the creation of gardening aprons from compost bags. These are quick and easy to make. Having cleaned the bag, cut them into rectangles and attach cotton tape or cord to hold the apron in place.

Balls

Old tennis and golf balls can be fitted on top of stakes to prevent eye damage when weeding.

Bamboo

If you have bamboo growing in your garden, cut out the oldest canes from the middle of the plant each autumn. Store the canes in a shed or garage until spring, when they will be completely dry. These canes can be used to support other plants growing around the garden. Any wiry side branches can also be cut and dried, since these are perfect for supporting tall-growing perennials.

Barrels

Old barrels from breweries make extremely good containers. Half barrels can be dug into the ground as small water features. The wood swells naturally and holds the water in place, although water levels will need topping up from time to time. Place an old brick or stones at one side of the barrel to provide an exit for any hedgehogs that may fall into the water.

A vintage bike cum planter creates a decorative garden feature.

Bikes

Painted up and turned into a decorative feature, old bicycles make an eye-catching item in the garden. The seat and basket are frequently replaced with planters, and small pots hung from the handlebars.

Biodegradable packaging

Starch-based packaging bearing a logo indicating that it is biodegradable can be placed on the compost heap. This material will decompose more quickly if it has been broken or cut into small pieces.

Blankets

Often used to cover compost heaps as a form of insulation.

Bras

Bras have been used to support developing fruit such as melons growing in the greenhouse.

Bricks

Old bricks are a very valuable resource around a garden. Any remaining mortar should be removed before reusing. Bricks can be reused to make raised beds or paths, and can be turned into edging for borders. Placed in small piles, old bricks make good supports for water butts. They are also useful in the vegetable patch as barriers around carrot crops. Carrot fly stays close to the ground – placing a single layer of bricks around the vulnerable crops may be a sufficient deterrent. Old bricks can be used to keep the edges of fleece, polythene and netting in place when protecting crops from birds or cold weather.

Broken crockery

Pieces of broken mugs, cups, plates and vases can be used to improve drainage for container plants. Put a few pieces at the bottom of each pot.

Many gardeners often reuse broken crockery to create pretty mosaics on walls or within paths. The amount of pieces required will depend entirely on the size of the project. Having decided on a design, draw a full-size plan on paper or old cardboard. Lay the broken pieces on the plan so that you can see exactly where each piece will go. This will enable you to see exactly what the end result will look like. When you are satisfied with the final design, spread some cement onto the chosen location. Following the outlines of your design plan, place the pieces of broken crockery firmly into the cement. Leave to dry.

This mosaic technique can also be used to decorate pots and other containers to give them a totally different appearance.

Brushwood

Use stems cut when pruning trees and large shrubs to create fences hiding compost bins or rubbish bins. Weave pieces in and out of vertical stakes. This method results in ideal temporary windbreaks as well as being eco-friendly since they will provide shelter for birds and small mammals.

Brushwood stems can be used to create stakes to support tall-growing perennials. Cut them into appropriate sizes in the winter ready for use in the spring. Eventually when the stakes rot down, they can be added to the compost heap.

If you have an open fire or log burner, cut the brushwood into manageable sections and store under cover. Allow the brushwood to dry out thoroughly. This usually takes approximately twelve months. The material can then be used for kindling.

Bubble wrap

This provides good insulation for pot plants against frost. Wrap around the sides of pot plants to keep them warm. Bubble wrap can also be attached to the insides of unheated greenhouses to provide extra insulation.

Buckets

Broken buckets can be used to protect rhubarb or seedlings as well as helping to force early crops.

Building rubble

Any form of building work, from creating a driveway to building an extension, results in a lot of rubble. This can be reused as a base for new pathways, filling in holes in existing pathways or creating a base for a new shed, greenhouse or patio.

Some gardeners have created DIY rubble rocks for use in rockeries. These rocks are created by digging roughly shaped holes in the ground. The holes are then lined with plastic sheeting or old plastic bags. Pour in a layer of concrete and add some building rubble. Top up the hole with concrete, making sure that the rubble cannot be seen. Leave to dry and become solid. The 'rocks' can then be dug up and left to age naturally, although painting the rocks with layers of natural yoghurt will speed up the aging process.

Cable reels

Large cable reels left over from rolling out wire or piping have been upended and turned into garden tables.

Cans

Cans of any size can be reused as plant pots. Wash thoroughly, and drill a hole in the bottom for drainage. Cover the outside of the can in brightly coloured paint, fill with compost and plant. These can be arranged in groups on patios, or turned into pretty vertical planting along a drainpipe.

Cardboard

All forms of cardboard including cardboard tubes can be added to the compost heap. Cut it up into smaller pieces for quicker decomposition. Large pieces of cardboard can be used to provide insulation around the outside edges of a compost heap.

Another use of large pieces of cardboard is to turn it into mulch. Place the cardboard around vulnerable plants. Take care to weigh it down with stones or bricks otherwise it may be blown around the garden during high winds. The cardboard will generally survive to use as mulch throughout the growing season. By the end of the season, the cardboard will be breaking down and can then be dug into the soil, where it will decompose quickly and help provide nutrients.

Cardboard tubes

Kitchen roll and toilet roll tubes make extremely good seed planters for sweet peas, peas and beans. Tightly pack the tubes so that they are standing upright in a seed tray. Fill the cardboard tubes with compost and place one seed in each tube. When the roots are beginning to emerge out of the sides of the tube, the young plant can be replanted into the garden or into a larger pot as required. The cardboard tube will decompose naturally, so there is no need to disturb the plant roots when transplanting.

Kitchen roll tubes have been used for planting individual carrot seeds, especially by show gardeners. Filled with compost, the tubes allow the roots to grow long and straight.

Car parts

Once cleaned and free of any oil and grease, cars have been used as innovative, eye-catching planters. Typical examples have included filling the boot and engine compartment with compost and using the area to grow colourful, trailing plants covering much of the vehicle. Steering wheels have been removed and turned into mini herb beds, while removing the top of an exhaust box can result in an unusual container. Car seats have been covered with a layer of soil and used to grow thyme or camomile, thus creating a medieval-style seat.

During the Second World War, Wolseley Motors reused the windscreens of old cars as marrow and cucumber frames. Matthew Watkinson of Beeview Farm turned the interior of an old car into a natural greenhouse.

Turning cars into hen houses is another commonly used option. All it requires is the addition of a small ramp to allow the hens easy access to the interior. The resultant hen house can be easily secured at night, and provides plenty of warmth for the hens.

An old car turned into a home for chickens. (Angela Youngman)

Carpets

Old carpets make a good covering for compost heaps as they act as insulation.

An unwanted carpet can also make a good weed suppressant, especially within allotments or vegetable patches. Take care to brush off any soil that gathers on top of the carpet on a regular basis, otherwise weeds will grow on top.

Cartons

Cardboard milk and fruit juice cartons can be reused as plant pots. Cut off the top and use upright as containers for single plants. Alternatively, lay the cartons down. Remove the front pieces of the carton and use the remainder as seed trays. Always punch drainage holes in the base of the carton.

Cart wheels

Cart wheels have traditionally been used to create herb beds.

Chest freezers

Freezers have often been used on allotments as growing containers for plants such as melons that require warmth and space to sprawl. The freezer container is filled with a mix of compost, grass clippings and straw. There are safety considerations that must be borne in mind when reusing this type of container as there are inherent dangers. Lids must be removed so that there is no risk of anyone getting caught inside. Cooling systems and insulation contained within freezers pose another major issue. The Environment Agency classifies freezers as hazardous waste due to the fact that chlorofluorocarbon gases (CFCs), pentane and other chemicals can be found within the cooling systems and insulation. Although the majority of these chemicals are not directly harmful, the problem is that they react with the upper stratosphere and deplete the ozone layer, thus contributing to global warming. Removing these chemicals is a task that must be undertaken by licensed contractors.

Chicken wire

Keep pieces of old chicken wire. It is ideal for putting on top of newly planted pea, bean and sweetcorn seeds to prevent mice and birds from eating the seeds.

Chimney pots

Old chimney pots make extremely good containers, especially for invasive plants like mint or for trailing plants.

Chocolate box trays

Plastic chocolate box trays can be reused as seed modules.

Christmas trees

Natural Christmas trees can be shredded and turned into compost. Local councils operate special collection points where Christmas trees can be taken for shredding.

Cobbles

Widely used in many driveways and paths, these can be recycled to create new paths when undertaking landscaping projects.

Coffee filter papers

These can be reused as mulch around the interior of hanging baskets and planters due to their absorbent quality. The papers can also be added to the compost.

Coffee grounds

Coffee grounds are a useful compost material. Gardeners have also been known to advocate their use as slug deterrents or for placing around acid-loving plants like rhododendrons and azaleas. However, the RHS say that it is unlikely that coffee grounds will retain enough caffeine to control slugs, nor would they be sufficiently acid to alter soil pH, but they are unlikely to harm acid-loving or ericaceous plants.

The RHS say that coffee grounds are a useful source of organic matter when added in moderation to the soil or compost bin. Freshly ground coffee remains should not be used since caffeine is 'reportedly harmful to valuable soil organisms such as earthworms'.

The general consensus of opinion among plant organisations is that coffee grounds are best used in compost, but small quantities can be sprinkled onto the soil and dug in.

Colanders

Attaching wire supports and filling with suitable planting materials will turn a colander into a hanging basket.

Upturned metal colanders have been turned into unusual light fittings when combined with a cool bulb.

Comfrey

A pretty perennial herb, it looks lovely in the garden. The leaves can be reused to create an extremely nutritious natural fertiliser, containing high levels of potassium, phosphorus and nitrogen. Pick the leaves and place in a bag. Punch holes in the bottom. Tie the top of the bag tightly. Put the bag into a container of water – an old water butt is ideal. Add a tight-fitting lid. Leave for about two weeks and then begin to drain off the liquid fertiliser. Make sure it is well diluted – about fifteen parts water to one part comfrey liquid. Use on all plants within the garden. The comfrey liquid will last a long time as long as it is stored in a cool place.

Compact discs

Bright and shiny, they can be used to scare birds away from newly planted crops or seeds. Tie the compact discs onto stakes or poles so that they move around in the

A compact disc used as a bird and insect scarer.

wind. If fastened to securely to stakes, they will also make good reflectors along a dark drive or path.

Compost bins

Damaged compost bins can be converted into tree guards.

Concrete

Broken concrete blocks and paving can be reused as hard core when putting up garden buildings or creating patios.

Half-used bags of concrete mix can be used up by creating a series of bespoke paving stones. Pour out a selection of circular sections, then use a metal rod to draw shapes on the surface before leaving to set. This results in some very decorative stepping stones that can be used anywhere in the garden.

Containers

Used ice cream containers can be reused when planting out seeds or cut up to make plant protectors against rabbits or slugs.

Cooking oil

Combined with a large container of sand, it makes a good cleaner for tools such as spades and forks. Some gardeners have reported diluting used cooking oil with paraffin and using it as a conditioner for wooden garden furniture, compost bins and fences.

Corks

These are perfect for placing on top of garden canes. They will protect your eyes from the sharp spikes.

Corrugated cardboard

Small pieces can be scrunched up and added to the compost or used to keep seed potatoes in place when chitting.

Corrugated metal

Combined with wood slats and overlapping pieces of wire netting, rusted corrugated metal creates ideal protection for young plants that could easily be removed and

A home-made protective cage protecting brassicas using corrugated metal, wood and netting. (Angela Youngman)

used elsewhere when necessary. Pieces of old corrugated metal are frequently used on allotments and vegetable areas as they create ideal rabbit-proof fencing.

Corrugated plastic

Turn lengths of corrugated plastic into simple cloches to protect a row of plants. Secure the plastic at the sides and ends, and block the ends at night with something flat such as a piece of wood. Bear in mind that the plastic will deteriorate over time and leave plastic debris in your soil. As soon as any evidence appears that this is happening, you should immediately remove all the remaining plastic and dispose of it carefully.

Cutlery

Old spoons make ideal tools for use when transplanting tiny seedlings.

Cycle helmets

Passing a doorway in a market town revealed the unexpected sight of hanging baskets made out of cycle helmets. A few holes drilled into the bottom to provide drainage, plus the addition of compost, resulted in some very unusual, and attractive, lightweight baskets.

Dishes

Shallow dishes can be reused as bird feeders and bird baths.

Dustbins

Old or damaged dustbins can be turned into plant containers. Put some holes in the bottom to provide drainage. Fill with straw, grass clippings, compost and use to grow deep-rooted vegetables such as carrots, parsnips and potatoes. If using to grow potatoes, fill only half the bin with compost before planting the tubers. Then add the tubers and a layer of compost. As the plants grow, more compost can be added in order to earth up the stems. This will encourage more tubers to develop.

Dustbins can be turned into tree guards. If the tree is small, the container can simply be placed over it so that it surrounds the trunk. A larger tree would need the container to be split open before being placed around the trunk. Since the containers are quite large and dense, they prevent rabbits and deer from munching on the young tree's bark, thus allowing it to establish itself.

Egg boxes

Cardboard egg boxes can be broken up and added to the compost heap. They can also be used as planting containers for seedlings. If required, the egg box can be planted straight into the garden or into a larger container since it will decompose naturally over time. This will avoid the risk of any root disturbance for the young seedlings.

Reusing egg boxes as planters for seedlings.

Egg boxes and trays are useful containers for sprouting potato tubers. Individual tubers can be placed safely in separate sections, allowing air to circulate and providing space for the eyes to sprout.

Egg boxes can also be used to store fruit. The individual units ensure that fruit is held securely without touching other pieces, slowing down the speed of ripening. There is less chance of fruit becoming bruised when stored in this way.

Egg shells

These are a popular traditional method for dealing with slugs and snails. Collect up empty egg shells and break them into small pieces. Placed around vulnerable plants, they can act as a natural deterrent.

Fabrics

Natural fabrics such as cotton and wool will break down naturally. Cut up and add to the compost. Unwanted and damaged wool blankets can be used to provide warm coverings for compost heaps. Place thick pieces of fabric inside hanging baskets to help retain moisture.

Feathers

Feathers can be added to compost and left to degrade.

Fencing

Wooden fence posts can be reused if undamaged, while posts unsuitable for reuse can be cut to use as lawn or path edging, or as vegetable and fruit supports.

Reusing trellis panels as a free-standing fence. (Marie Shawcross)

Fence panels broken down in storms or removed due to a change in garden design can often be salvaged by undertaking some repairs, cutting down to create a smaller fence around another garden feature, or even turning into a trellis against the new fence.

Landscape designer Marie Shawcross reused trellis panels from a re-designed front garden. Measuring 6ft by 6ft in a large diamond pattern, and made in good-quality planed wood, the panels had provided a brilliant support for climbers on a free-standing fence. Reusing the trellis as a climber support attached to a feather-edge fence provided solid support for heavier climbers such as roses and grape vines since it allowed the air to circulate, thus reducing the likelihood of pests and diseases. When the fence needs a coating of preservative, the panels can be unhooked with the plants still attached and laid on the ground.

Flour shakers

These are an excellent way of sowing small seeds. Combine the seeds with a small amount of fine sand and shake across the planting area. It ensures better seed dispersal while lessening the risk of the seeds sticking together or clumping.

Foil

Cooking foil can be reused for mulching purposes. Wash it clean of any residues before using. Place it around crops like peppers, tomatoes and chillis that require extra heat. Make sure the soil is warm and damp before covering with mulch. Leave room around the plant stem to allow for watering.

Foil tops from yoghurt cartons or milk bottles

Collect up a large quantity and thread them on string. Place stakes at either end of a newly planted area and stretch the strings across the beds in order to deter the birds from taking the seeds.

Fruit punnets

Fruit punnets make lovely seed trays. Lidded versions are natural mini greenhouses since closing the lid will hold in moisture and heat.

Gabions

Gabions are rectangular cage-like structures used by commercial organisations to transport large quantities of materials. Filled with large stones and rocks, gabions have been used by the military and structural engineers for centuries as a way of protecting river banks and shorelines from erosion. In recent years, they have become a popular material landscaping system among garden designers and landscapers. Laid at angles or stacked like large bricks, gabions can be turned into benches, sound breaks, walls or boundary markers around a garden. Landscapers have filled the gabions with all kinds of material including shells, wine bottles, toy cars, trucks, rubber balls and steel globes as well as the more conventional cobbles and rocks.

A wall created out of gabions. (Carol Whitehead)

When creating a gabion-based structure, it is important to ensure stability by making it wider at the bottom, with lighter materials used to fill the upper gabions. Matthew Watkinson of Beeview Farm, Pembrokeshire, has used gabions in many other ways such as turning them into log stores, a biodigester and a reed bed system to clarify grey water. Gabions are widely available for purchase online, while smaller ones can be constructed out of strong leftover wire mesh.

Glass jars

Most kitchen cupboards will reveal a large quantity of glass jars filled with chutneys, jams, cooking sauces, pasta sauces and preserved vegetables. Wash well and reuse the jars for storage. They are ideal for storing seed packets, screws, seed labels and other small items. At harvest time, glass jars topped with screw top lids are ideal for storing surplus produce from the vegetable garden that has been turned into jams and chutneys.

Tea lights placed inside small glass jars provide pretty illumination for evening parties and barbeques.

Larger jars can be turned into unusual mini gardens housing small delicate house plants. A thick layer of gravel is required at the bottom of the jar, and then compost to which some charcoal has been added. Old spoons are ideal when planting or maintaining the garden. Periodically cover the top of the jar with a plate or lid.

Moisture released from the soil and the plants themselves will be captured on the sides of the jar, before running down to re-water the plants.

Grapefruit skins

Leave on the ground overnight and slugs will be attracted into the skins. Coconut shells, bricks and pieces of slate can also act as slug attractants.

Grow bags

Depending on the initial crop grown within a grow bag, it may be possible to reuse the compost for a quick catch crop of lettuce or radishes. Much of the nutritional content of the grow bag will have been utilised by the original crop. Ultimately the compost from a grow bag can be dug into garden borders or a vegetable plot as a soil improver.

Guttering

Guttering makes useful plants or supports. Gardeners frequently use it as a way of growing cucumber plants so as to avoid getting water on the stems. Alternatively, it can be used for growing seedlings like beetroot and radish that dislike experiencing any root disturbance. When the seedlings are large enough, they can be slid into a pre-prepared trench within the vegetable area. Placed upright, pieces of guttering will act as supports for nets, or materials to create grow tunnels. Fixed firmly against a wall or fence, strips of guttering can result in some pretty vertical gardening displays when pots are inserted down the strip.

Hats

Unwanted hats can be turned into pretty hanging baskets or planters, but always remember to add a plastic liner (an old plastic bag or sack will serve).

Turning a hat into a planter. (Colton Care)

Hollow plant stems

A traditional Victorian gardening practice was to place sections of hollow plant stems horizontally in fruit trees. Earwigs and other pests automatically gravitate into the stems, using them as shelter. The stems and pests could then be removed safely.

Hot water bottles

These can make good kneelers. Fill the hot water with sawdust, foam chips or sand. Cover one side with some old carpet. Use to kneel on whenever the ground is soggy, or there are sharp stones. It will make the task more comfortable and protect the knees.

Cut up a hot water bottle and use as slip mats for plant pots.

Cut into even smaller strips and place at the bottom of plant pots to improve drainage.

Ice cream containers

Old ice cream containers can be turned into useful slug traps. Cut some holes close to the top of the sides of the container. These holes should be big enough to allow slugs to enter. Half fill the container with beer or milk. Fasten the lid on tightly, and then bury it in the soil so that the holes are level with the ground. The beer or milk will attract the slugs, which then drown in the liquid. The only drawback is the need to empty the container on a regular basis.

Ice cream containers are good for storage, or as containers for collecting produce from the vegetable patch.

Ironing boards

The metal tops of ironing boards can be laid out on the ground to form a path. Covering the metal with a rust converting paint such as Hammerite will ensure the path remains useable for a long time. They are best used with a landscaping material designed to prevent weeds from growing underneath.

Jiffy bags

These can be added to the compost bin. It is a messy job cutting them up, but it will ensure rapid decomposition.

Kitchen paper

When used, add to the compost mix. It will break down very quickly.

Ladders

Old ladders can be propped against a wall and used as a planter. Plants can grow up the sides of the ladder, while the steps can be used to hold pots at different levels creating a pretty display feature.

Logs and branches

Create a log pile in a corner of the garden using woody prunings and large pieces of wood. The material will slowly rot down and can eventually be used on the compost heap. In the meantime, it will act as a home to frogs, toads and beetles – all of which are beneficial predators dealing with insect pests preying on garden plants.

Large logs can be used for lawn or path edging, or as fence posts. Branches can be turned into home-made trellis.

A home-made trellis using pruned branches. (Angela Youngman)

Lollipop sticks

These sticks make excellent seed markers.

Logs

Long lengths of logs from the trunks of cut down trees make ideal natural edging materials for flower borders and paths. Logs of any size are good wildlife habitats when left in a corner of the garden. Another very large log was turned into a stand for use when chopping kindling.

Louvre panels

Four large louvre panels can be combined to create a simple shed ideal for holding a selection of garden tools. Just add a roof and a shelf.

Manure

Horse manure is a traditional fertiliser that has been used for centuries. It is one of the best types for use in the garden as it is fibrous and contains approximately 0.6% nitrogen. It is ideal for adding to compost heaps, creating hot beds or dug into heavy

Manure – a valuable addition to any garden.

soil the autumn before planting takes place. It is essential to allow horse manure to mature before using.

Plants can be scorched if fresh manure is used around them. For quicker use, it is possible to take half a bucket of 'neat' horse dung and fill to the top with water. After a couple of days, a cupful of the resultant natural fertiliser can be added to a watering can, diluted and used to water vegetables, tomatoes, roses and other nutrient hungry plants. The bucket should be kept topped up with water. When the growing season comes to an end, add the remaining contents to the compost heap.

Cow manure contains around 0.4% nitrogen. It can be added to compost heaps or mixed directly with sandy soils some months prior to planting. Cow manure is extremely liquid so needs to be mixed with an absorbent material such as straw before using.

Poultry manure is rich in nitrogen – about 1.8%. It can be very messy to handle, and needs to be combined with an absorbent material. This type of manure is ideal for adding to fallow ground.

Rabbit manure is rich in nitrogen and should be added to the compost.

Pig manure is equally nutritious and should be used in the same way as horse manure.

Pigeon droppings from pigeon lofts are nitrogen-rich and can be added to compost.

Goat manure should be used in the same way as horse manure.

Sheep manure is very high in nutrients and is best used to make a liquid manure. Half a sack of sheep droppings will make enough liquid manure for use on a medium-sized garden for a year. To do this, you will need a sack, stake and a large metal or plastic watertight container. Fill the container with water and half fill a sack with manure. Tie the top of the sack tightly with string. Loop the string over a wood stake and place the stake over the top of the container, allowing the sack to just rest in the water. After about two weeks, the water will have turned a rich brown colour. Remove the sack from the water and empty the contents onto the compost heap. Cover the container and use the diluted liquid on the garden.

Metal edging

Strips of aluminium, steel or other metals can be turned into very durable border edging.

Milk cartons

Gardeners have reused milk cartons into compost scoops. The cartons can also be cut into strips with a pointed end for placing in the ground when using as a plant label. At the end of the season, nail polish remover can be used to erase whatever has been written on the label so that it can be reused.

Milk residues left in the bottom of milk cartons, or stale milk can be diluted with water to use as a spray to get rid of plant moulds.

Mortar

Mortar from around bricks can be removed and broken up. This can be dug into heavy soil as a conditioner. Make sure there is no rubble attached when digging it in.

Moss

Moss can be a problem in lawns and can grow very quickly. Gardeners often pull it away from the earth and spread it on a tray. When it has darkened in colour and is dry to the touch, it can be added to hanging baskets as a natural aid to water retention.

Net curtains

Net curtains can be reused as crop protection, especially against carrot fly.

Nettles

An invasive plant but can be reused usefully. Shred the green stems and leaves of stinging nettles before they flower and begin to seed. These make a nutritious addition to compost heaps. In spring, the young leaves can be harvested carefully and used to make nettle soup.

Nettles are invasive but can be useful!

Newspaper

Newspaper can be used as mulch within the garden. A very thick layer is needed to suppress weeds. It can look a bit messy, so should be used on vegetable plots or at the very back of borders. Use heavy stones or bricks to hold the newspaper in place. Over time, the paper will break down and can then be dug into the soil as a conditioner.

Placing one or two sheets of newspaper in the bottom of seed trays before adding potting compost will help maintain moisture levels.

Newspaper can be turned into useful pots for planting seeds. Fold pages in half lengthwise and then roll it around a cardboard tube, leaving part of the paper sticking out. Gather up the loose ends at the bottom and fold inwards to create the bottom of the pot. Slide out the cardboard tube and flatten the bottom by filling the pot with compost. Place the pots tightly together in a tray for additional support. When roots begin to emerge through the paper, the pots can be planted into the ground without causing any root disturbance. Smaller pots can be made by cutting long strips of newspaper, before winding around a cardboard tube.

Newspaper can also provide extremely good emergency anti-frost cover. Spread the papers over the plants and peg down with stones or pieces of wood.

Creating pots out of strips of newspaper. (Angela Youngman)

Oil drums and barrels

Large metal drums are often used to transport large quantities of liquids such as oils. These drums have been turned into simple barbeques by cutting them in half horizontally, and adding hinges and a handle. The drum barbeque was placed on two pillars of old bricks, with shaped bricks on either side of the drum to hold it securely in place. Strong wire mesh inserts were used to hold the charcoal, and a grill was added for the food. To give it a decorative effect, the bricks and drum were painted in co-ordinating colours. At Antiques by Design, an online business that reclaims antiques and garden items to create innovative products – especially lighting and mirrors – a metal barrel was cut up and turned into a mirror. Other uses include DIY incinerators and firepits.

A mirror made from the bottom of a water barrel by Guy Chenevix-Trench. (Antiques by Design)

Orange nets

Oranges are often purchased in nets. Open carefully and then reuse the nets to store onions over winter.

Paintbrushes

Children's old paintbrushes can be washed and then used to transfer pollen when growing plants like aubergines and pumpkins.

Pallets

Pallets can be reused as storage, or upcycled to make planting areas. They make useful cold frames or mini greenhouses by adding old windows to the sides or roof. Pallets also make good shelves within greenhouses as long as they are fixed securely in place. Many gardeners combine several pallets to create home-made compost bins, benches, tables and other garden features.

Pallets have been used to create simple fences. They can be fixed directly to posts dug into the ground so as to mark boundaries. Such fences can be made more attractive by painting the surfaces, or attaching decorative features.

DIY garden furniture made out of pallets.

Paper

Paper will break down in the compost heap. Shred or cut up large sheets of paper, bills and confidential documents before adding to the mixture.

Pet bedding

Bedding from vegetarian pets such as rabbits, hamsters and guinea pigs can be added to the compost.

Plastic bags

Plastic carrier bags can be reused as bird scarers. Tie the bags securely to a tall stake or post and leave them to flap in the wind.

Plastic drinks bottles

Plastic drinks bottles can be reused in many ways around the garden.

They make useful watering containers. Pierce small holes in the bottom and sides of the bottle. Dig a hole the same size as the bottle beside a vulnerable plant. Fill the bottle with water and refasten the cap on the bottle. Refill the bottle as needed. The water will filter steadily through to the roots of the plant.

A plastic bottle turned into a mini cloche. (Angela Youngman)

Bottles can be turned into self-watering containers. Large bottles are most effective for this task. Cut into two halves around the middle of the bottle. The bottom part should be filled with water. Take the top half of the bottle and turn it upside down. Place it on top of the water so that the neck part is resting in the water. Place a wick or a piece of absorbent material such as capillary matting through the neck of the bottle so that it lies in the water. Fill the top part of the bottle with compost and a plant. Refill the bottom of the container with water whenever the water level drops.

Plastic bottles make excellent mini cloches. Cut out the bottom or side of the bottle and use to cover vulnerable plants whenever frosts are forecast or to encourage strong seedling growth within the vegetable patch. It will also act as a deterrent to mice, rabbits and pigeons from nibbling young shoots.

Placed on the end of canes or stakes, plastic bottles can be used to hold up netting or fleeces so as to make a temporary fruit or vegetable cage. The stakes should be taller than the plants that are being protected. Put a plastic bottle on the end of each pole, before throwing the netting or fleece over them. Weigh down the sides of the netting where it touches the ground with stones, bricks or large pieces of wood.

Plastic bottles make good bird feeders. Pierce holes in the side so that birds can get to the food. Put sticks through the bottom holes to give birds a place to perch, while taking food from the holes above. Use string around the neck of the bottle to hang the feeder from a tree – but make sure that the top is screwed on safely to ensure that small birds are not trapped inside the bottle.

Plastic pipe

Long lengths of plastic pipe are ideal for making simple arches that can be stabilised with wood or metal stakes. Such arches are ideal for acting as a support for beans,

Cover plastic pipe arches to create a polytunnel for growing plants.

tomatoes as well as climbers like hops and honeysuckle. Flexible plastic pipe can be used to construct polytunnels.

Plastic storage boxes with clear lids

Plastic storage boxes with clear lids can be used as unheated propagators. Place drainage holes in the bottom of the box. Fill with compost and add seeds. Take care to leave space between the compost and the lid. Water the compost gently and place the clear lid on top. It will encourage germination and maintain humidity. When the seedlings begin to emerge, remove the lid.

Plastic trays

Plastic trays used by supermarkets to package meat, fish, vegetables and fruit can be reused as seed trays.

Plastic ties

Reuse plastic ties or wire closures. They are ideal for fastening bags and containers when storing produce, seeds or for attaching plants to stakes.

Plastic sacks

These can be reused to line hanging baskets.

Plastic pots

Fill small plastic pots such as yoghurt pots with home-made fat balls containing bird seed. Allow the fat balls to set solid, before turning upside down and attaching to trees or a bird table.

Polystyrene packaging

Apart from being reused as planters, large pieces of polystyrene packaging are useful for placing under developing fruit such as melons, pumpkins or squashes. The insulation qualities provided by the polystyrene helps fruit development. It also ensures the young fruit rests on a smooth, clean and dry surface unlikely to cause bruising.

In spring, a level piece of polystyrene material placed under seed trays will encourage germination. The polystyrene keeps the base of the tray warm.

Individual polystyrene cups can be reused for planting marrow and tomato seeds.

Reusing polystyrene packaging to create insulated planting boxes.

Small, broken pieces of polystyrene can be placed in the bottom of plant pots in order to aid moisture retention.

Pots

Ring culture pots can be easily created out of damaged plant pots. The pots should be approximately 17cm (7in) in diameter. Remove the bottom of the pots. Place the pot on the growbag and cut out a matching circle. Push the pot into the growbag. Fill up the pot with compost and add a tomato plant. The extra compost in the pot provides a more extensive rooting area, and helps prevent the plant from drying out in the summer. Watering directly into the pot reduces water wastage.

Pumpkin shells

Rather than being thrown away after Halloween, pumpkin shells can be cut up and buried in the ground, or placed in the compost. Save the seeds for planting next year, or alternatively roast them for eating or feeding the birds.

Railway sleepers

Safety considerations mean that railway sleepers have to be replaced at regular intervals. These have become a traditional form of garden edging, or are laid flat to create paths and steps.

Rotary washing lines

Remove the washing line and turn the pole upside down. It creates a very sturdy climbing frame for runner beans. Additional supports can be created using long pieces of washing line. Tie long lengths of line stretching from the top of the pole to the ground. Fan these out like a maypole and fix securely in place.

Finding a new use for an old rotary washing line as a bean support. (Angela Youngman)

RECYCLING IN THE GARDEN

Rusty pipes

At the 2018 RHS Tatton show, Bees Gardens used rusty pipes as vertical structural elements within a shady show garden designed to raise awareness of the Stroke Association. Weaving through the garden were two lines of rusted metal tubes providing vertical accents and these represented the cerebral artery most commonly affected by strokes. Scrap and disused gas and drainage pipes were used to add height to the front border.

Scaffold boards and railway sleepers

Unwanted scaffold boards and old railway sleepers make great benches, raised beds and fence posts.

Sand pits

When no longer required, children's sand pits are frequently repurposed as small pools and water features.

Sanitary ware

Sinks, baths and toilets are often reused around the garden. Sinks and toilets have been used as containers, particularly for invasive plants. Sinks are often used as troughs for alpine plants. They can be given the appearance of age by covering

Flowers growing out of an old toilet.

the exterior with concrete. Adding a layer of yoghurt to the exterior surface will encourage mould to grow.

Baths make excellent small ponds. Dig a hole slightly larger than the bath. Fill the plughole with concrete to prevent water draining away. Cover the base and sides with dark plastic sheeting and fill with water. Add some bricks at one end to create a shallow edge so that wildlife can climb out.

Seaweed

Children tend to collect this in buckets whenever they go to the seaside, take it home and then forget about it. Seaweed can be added to compost heaps, or placed around roses as a combination mulch and fertiliser.

Shells

If you have seafood such as mussels, the shells can be reused. They can be used to provide drainage at the bottom of pots, or can be broken into pieces and used to create an attractive mulch around container plants. They will eventually degrade, providing useful nutrition to the soil. Many gardeners believe that placing a layer of broken shells around plants such as hostas will deter slugs and snails from attacking the plants.

Shoes and rubber boots

Unwanted footwear such as shoes and rubber boots can be turned into very decorative planters by adding drainage holes, compost and plants.

Boots become pretty planters.

Socks

Old socks are extremely useful as soft plant ties.

Soil sacks

Every gardener inevitably buys some bags of soil conditioners, bark mulch or planting compost at some point. The sturdy black plastic bags can be reused very effectively to grow potatoes and carrots. If growing carrots, half fill the bag with compost and add the seeds. Fold down the sides to create the height required. This method will reduce the problems caused by carrot fly, and encourage the development of long roots. For potatoes, it is best to start with a layer of compost at the bottom. Place the tubers on top and cover with compost. Add more layers of compost as the plants grow, at the same time pulling up the sides of the bag to provide a dark environment for the developing potatoes as well as a container. Keep going until you reach the top of the bag. This generally involves at least three layers of compost. When harvesting, simply turn out the contents onto a layer of newspaper.

Steel plate sections

Sometimes described as steel skeletons, steel plate sections comprise leftover material from metal-cutting businesses, which may have been fulfilling an order for specific shapes to be cut out of the steel. Often available from salvage yards, these leftover

A decorative fence made from metal shape sections.

steel plate sections can provide useful decorative features for use in paving, screens, fences and railing. The sections can be painted using a rust-converting paint, or metal paints available in a variety of colours. Alternatively, the sections can be left to weather naturally, turning a deep rich brown.

Sponges

Sponges used for washing up or cleaning provide useful water storage at the base of pots or between pots and saucers. When watering during dry spells, remember to allow for the fact that the sponges do hold a large quantity of water, so you will need to water for longer. The main consideration to be borne in mind when reusing sponges within the garden is to make sure they are fully clean. Wash thoroughly to remove any traces of cleaning agents that might be toxic to plants.

Steel mesh

Concrete is frequently reinforced by an interior steel mesh, which is often cleaned up by gardeners and used to create a simple trellis.

Straw bales

Straw bales can be reused to provide the sides of cold frames, with the addition of old windows to form a roof. The straw bales provide natural insulation and warmth, and are quite sturdy.

Another use for straw bales is to turn them into raised beds, using them to create a base and sides. Use a weed-suppressant fabric to line the raised bed, before adding compost.

Tea bags

Used tea bags are often added to the bottom of hanging baskets to aid water retention.

Teapots and jugs

Damaged teapots and jugs are frequently turned into decorative plant containers.

Terracotta pots

Inventive gardeners have been known to repurpose large terracotta pots for hose storage. A large pot is placed upside down on the ground. The hose is wound around it and the nozzle is pushed into the drainage hole to keep it in place. This type of storage method prevents the hose kinking.

Smaller terracotta pots have been used as hose guides around the garden. Two terracotta pots are placed on top of each other to form an hourglass shape, leaving the narrowest part in the middle. A dowel or stick is pushed through the centre of the pots and into the earth below to hold them securely in place. When watering, the hose can be allowed to drape around the pots, thus preventing it from being dragged over vulnerable plants.

Tights

Strong and flexible, sections of tights are frequently reused as plant or tree ties.

Timber

Timber of any kind, no matter how small, can be reused around a garden. Stakes are always needed to support trees or fences. Long lengths of timber can be placed sideways to make border edging or a raised bed. Smaller pieces can act as weights to hold down netting or fleeces. Children can happily play for hours in a timber den created by some old crates in a corner of the garden. Use an old tarpaulin on the ground to provide a floor.

Seek out recycled or reclaimed timber wherever possible when planning a new project as it is already seasoned and unlikely to warp.

Tin cans

Large tins are frequently reused as storage containers for seed packets and other small items around the garden. They make very attractive decorative containers, especially

Recycled tin cans used as flower vases and planters.

when planted with crops such as chillies, peppers, salad leaves and tomatoes. Before using as a planter, make sure that they have been cleaned thoroughly and add some drainage holes to the bottom of the can. A row of metal cans filled with plants can make very decorative border edging along a patio.

Tools

Old tools can be turned into decorations, such as a handle for a shed door. Sometimes tools can be given a new function. One manufacturer, Niwaki, has taken traditional Japanese tool designs and given them new uses within the domestic garden environment such as Niwaki's herbaceous sickle. In Japan, this tool is used for harvesting rice but it works equally well on grasses and herbaceous plants within the garden.

Toolboxes

Old toolboxes make distinctive and very sturdy planters. Tool box planters are very stable and less likely to tip over by accident. Drill a handful of drainage holes in the bottom, then fill with compost together with a selection of low-growing plants and succulents.

Toothbrushes

These are ideal for cleaning in small spaces or for awkward items like the teeth of saws and chainsaws.

Tree branches/tree trunks

These can be cut in slices to create stepping stones, while thinner sections are frequently used to form border edging.

Trugs

The handle might be damaged, but the base can still be used as a planter housing a few annuals or salad crops.

Tyres

These are frequently used to make plant containers due to the fact that they are moisture-retentive and durable. Clean them thoroughly before use. Depending on the height of the planter required, place one or more tyres to create a small pile. Line the interior with an impermeable material to prevent toxins escaping from the tyres. Fill with compost. For a decorative effect, paint the outside of the resultant planter. Tyres are extremely useful as a way of growing potatoes. Begin with one tyre and when

Used tyres turned into decorative border edging, which can also act as a play item for children.

the leaves begin to show, add another tyre on top of the original one. Top up with compost. Repeat the process at least once more during the growing season. No digging is required. When it is time to harvest, simply remove the tyres to give access to the tubers. The compost can be reused elsewhere in the garden.

Compost bins can be created out of tyres using the same method. Place two or three tyres on top of each other and fill with compost materials. Cover with a thick piece of carpet.

Many gardeners have also created tables out of two or three tyres. Use industrial superglue to fasten the tyres together, and paint your desired colour. Add a table top made from surplus wood. A similar technique can be used to create seats around the garden.

Another frequently seen method of reusing tyres is to turn them into border edging, a balancing installation for a play area, as a swing or as boundary markers. Collect a group of tyres and bury each one halfway down in the ground. Depending on your inclination, the tyres can be varied in size, or linked together.

Umbrellas

Umbrellas all too frequently get broken, especially when they are blown inside out by gusts of wind. Rather than throwing them away, gardeners have put them to many

Creative use of an old umbrella.

alternative uses. Attached to fences, and filled with fake or real flowers in little pots hidden among the folds, umbrellas can make pretty decorative objects. Turned upside down and filled with compost, they can form planters for eye-catching temporary summer displays.

Alternatively, umbrellas can be used to provide shelter from hot sunlight over sensitive plants. Transparent umbrellas are often turned into temporary cloches to protect young seedlings from early frosts. Removing the fabric shell and stretching out the framework can create a lightweight growing frame for peas and sweet peas.

Used compost

Growing materials used in planters, or from growing potatoes and beans in containers, can be recycled by digging into the garden soil or mixing with some fresh compost. Avoid reusing compost that has housed diseased plants or contains soil-borne pests.

Spent compost is also useful as a way of storing winter vegetables after harvesting. Root crops can be kept cool and in good condition for longer if stored in such compost within old packing crates, wine crates or discarded drawers as it forms a version of a traditional root clamp used for generations by farmers.

Vintage objects

Vintage objects such as bird cages, metal kettles, watering cans and metal crates are often used to create decorative planters or as a way of displaying a variety of small pots. Old tools can be turned into decorative lights.

Garden shears converted into a lamp. (Antiques by Design)

Washing machine rotary drums

These have been turned into fire pits. Always check that any plastic inserts or surrounding plastic material are removed completely, as this can be toxic when heated. If the drum has a spindle, this can be used as a support since it can be pushed into the ground to reduce the risk of the fire pit tipping over.

Water butts

Damaged water butts can be reused as tree guards against damage by rabbits and deer.

Damaged water butt turned into tree protection against deer and rabbits. (Angela Youngman)

Wellington boots

Punch some holes in the bottom and turn them into decorative plant pots.

Wheelbarrows

Unwanted or damaged wheelbarrows are frequently turned into mobile planters. Metal wheelbarrows can also be used as temporary incinerators to burn waste.

Wheelie bins

Damaged wheelie bins are highly prized on allotments. They can be turned into mobile water butts and compost bins, as well as providing mobile storage for tools, plant pots, stakes and fleeces.

Whippy prunings

These make ideal frameworks for crop protection or can be linked loosely together to create longer supports for crops such as peas.

Windows

Old windows are frequently recycled as covers for home-made cold frames. Double-glazed windows are best as they offer extra insulation against frost.

Old windows being reused to cover a cold frame.

Enterprising gardeners have been known to combine several windows to create a simple small greenhouse, adding a sheet of polycarbonate for the roof.

Join four small window panels together to form a box and add an inflammable base such as a tile. Place tealights inside to create a simple lamp feature. Leaving the top open will provide access and release any smoke during use.

Wire netting

Small pieces of wire netting are good for placing over a seed container or on the ground to create a template through which large seeds can be evenly sown.

Wine bottles

Turned upside down and dug into the earth, wine bottles have been used to make path and patio edging. Norfolk garden designer Rajul Shah used wine bottles to create a circle edging 3m across. The bases of these bottles were left standing about an inch above the ground. Mixing colours such as red, blue, white, green can result in a very pleasing and hard-wearing effect.

Wine bottles used as decorative edging by landscape designer Rajul Shah.

Wire from spiral-bound notebooks

Gardeners have used this spiral wire when staking plants, especially within the vegetable patch. The spiral is simply slid over the top of a cane and the plant stem.

Wooden crates

Lined with pond liner, crates can be usefully recycled as a base for smaller plant pots and containers, especially on balconies and patios. The pond liner helps retain moisture, thus making watering easier, and keeps the surface underneath clean. Most crates will hold several bags or pots.

They also make useful cold frames. If there is no lid, a sheet of transparent plastic or glass can be used, holding it in place with twine or string.

Yoghurt pots

These are one of the most frequently used recycled objects around the garden as there are many potential uses, the most popular being for storage or planting. Washed and dry, the pots can be used for storage, or turned into planters for seedlings. They can be cut up to create plant markers. Yoghurt pots have been used on top of canes to prevent accidental eye damage within vegetable plots, or as aids to hold netting and fleeces in place.

This list of potential recycling uses is not finite. Any conversation among gardeners, or a visit to a gardening group or allotment immediately results in countless other suggestions, proving that all that is needed when it comes to recycling and reusing is some imagination. Many of the ideas suggested above are the result of conversations in garden clubs and forums.

It must be said that there are some products that should not be recycled such as pesticides, bleach, anti-freeze and engine oil. Such items should be taken to a recycling centre for safe disposal.

The biggest problem encountered with regard to recycling materials is one of space. It is all too easy to simply start keeping materials on the basis that they might be useful one day. This can result in your house and garden being overtaken by surplus materials and incurring the potential risk of it looking like a scrapyard, resulting in some very unhappy neighbours. It is important to be ruthless – only keep what you really need for a specific project and use it. Pile tidily or keep out of sight.

Planning and organisation are essential when undertaking material recycling. Think about what you are doing in the garden, and what you are planning for the future. Look at potential materials that are being thrown out and ask yourself – would they have any use in the garden? Can they be recycled or can they be considered as a potential upcycling project? Always store any materials being kept for future use safely and neatly.

Repair, Reuse, Recycle

U ntil the 1950s, repair and reuse was the norm. Products such as lawnmowers, machinery, tools, greenhouses and garden buildings were purchased on the basis that they would last a long time. When they broke or became damaged, they were repaired rather than being thrown out. When a tool became blunt, gardeners reached for a sharpening tool and set about making it useable again. Throwing out garden items because they were no longer fashionable or because an improved version had become available was unthinkable. Machinery was used until it could be used no longer, and then broken down and reused in other ways. Make do and mend was a way of life.

In 1955, the American magazine *Life* published a story entitled 'Throwaway Living' that highlighted the way consumer and manufacturer attitudes were changing. It promoted the idea of single-use items as an essential part of modern life. The throwaway consumer society had arrived. Buy, buy, buy became the consumer's mantra. Products had a short lifespan and were just thrown out when no longer needed. Buying new was cheaper and more desirable than reusing or repairing. It was important to follow fashion and gardens became designer icons. Now the wheel has gone full circle, and the problems posed by declining natural resources are all too apparent. As a result, gardeners are regaining interest in methods of recycling, reusing and repairing.

It does take patience. It does not involve instant gratification in the same way as does going out and buying a new product. Time and effort is needed to repair an item or find alternative ways of using it. However, ultimately the effort is worth it.

Landscape and garden designers are increasingly being asked to create sustainable, eco-friendly gardens reusing as much existing material as possible or seeking out items that could be recycled for garden use. When asked to redesign a garden to create a better viewpoint to watch visiting badgers, garden designer Carol Whitehead was able to reuse most of the existing materials. A local arborist began the redesign by cutting back some of the boundary trees and shrubs, reducing but not removing the vegetation. A dead tree was turned into a 2.5m stump ideal for attracting perching birds, while the remainder of the trunk was cut up and left in piles to create a wildlife habitat. The lawn was replaced with a pond extending from the new patio layout and the remainder of the lawn was turned into a woodland-style planting scheme

A new pond and raised terrace created by reusing existing York stone. (Carol Whitehead)

designed to attract even more wildlife. A raised pond replaced existing steps built into the lawn, so as to encourage more birds to visit the garden. An upper terrace was created that enabled the home owners to watch the badgers prowling around at the back of the garden in the evenings. Existing York stone was recycled to make the new pond, a perch seat, steps and patio layout.

While undertaking another garden design project elsewhere, she discovered a marble bath being thrown out due to the presence of a hairline crack. By plugging the crack and plughole with mortar, the bath became a very pretty water feature filled with oxygenating plants that kept the water clean.

London-based designer Ed Oddy reused existing landscaping as part of a Kennington garden redesign. York stone had originally been laid in a haphazard way at the back of the garden, and the redesign enabled the blocks to be moved to form a dining area surrounded by new planting. To minimise waste, the bricks were laid directly onto an existing concrete slab and the top ground level retained as well as the existing concrete posts, while the fence was repaired rather than replaced.

Before buying anything new, it is important to take a close look at the old product and decide if it can be repaired or reused in some way. Sometimes, it just means buying a new handle or blade or getting the product serviced – much more cost-effective than buying new.

Spare parts can usually be obtained for most garden items. Manufacturers normally have to be contacted directly, or via their authorised distributors. The range of spare parts available is enormous, from covers for mini greenhouses to connections for solar lights and chainsaw blades. Local dealers will repair lawnmowers and other petrol or electrically powered tools. Such businesses can easily be identified via social media, newspapers, noticeboards or simply by asking neighbours.

Cracked glass in greenhouses can be replaced, likewise roofs and fittings. Increasingly, attention is being focused on repairing and moving rather than simply replacing. Moving a greenhouse or shed to a better position in the garden can be far cheaper than buying a new one. Jason Walker, of the Organic Garden Company in Teddington, south-west London, said,

> Greenhouses can be tricky to reconstruct but it is possible. There was one greenhouse which the original owner just wanted me to break up into a skip. I approached a second client who was happy to take it and even got the original owner some money for it too. Replacing the powder coating was very reasonable at £180 and with my labour, the new owner got what is a £900 greenhouse for less than £500.

Apart from breakages, the most common product problems experienced by gardeners are rust, damaged tools and punctures.

Metal objects such as gates, railings, arches, tools and wheelbarrows are likely to go rusty over a period of time due to the effects of weathering and regular use. Digging a flower bed can result in spades and forks being scratched by stone. Just a small scratch will damage the protective coating of any tool or metal object. The metal will eventually deteriorate and rust patches will emerge. If left untouched, the rust will spread and will eventually eat through the metal, creating holes that cannot be repaired. By taking swift action as soon as any evidence of rust appears, this situation can be prevented.

Use sandpaper to rub down rusty areas of wheelbarrows, gates and railings. This needs to be done carefully, removing all loose material. Wipe any residues and dust from the surface leaving a clean base. As long as there are no large holes, the surface can be painted with a rust-converting metal paint such as Hammerite. This paint changes any remaining rust into a harmless solid undercoat. When dry, cover the undercoat with two layers of exterior paint. Depending on the size of the task, it may take several hours to complete, but it does work out far less expensive than replacing the entire object.

Smaller rusty scratches on tools can be dealt with by using time-honoured methods employed by gardeners over the years that are now being rediscovered. One of the simplest ways of dealing with scratches is to fill a container with sand and used cooking oil. Place the metal part of a tool within the container and move it around. This will clean the metal surfaces and deal with any small scratches. The sand removes any residue and the oil prevents further rusting. Tools can be left standing in the container when not in use, so as to provide long-term maintenance.

A tool suffering from larger areas of rust can be cleaned effectively using an onion. Remove the loose skin and cut the bulb into large chunks that can be held comfortably in the hand. Sprinkle one side of a chunk with granulated sugar, and then use it to rub down the rusted metal part. The sugar adds friction and helps remove the rust. Change the piece of onion when it becomes dirty. Removing rust in this fashion does take time, but the tools do become very clean. It is a task best undertaken outdoors, as onions can make the eyes water. Always remember to wash your hands carefully afterwards to remove any residue as well as the onion aroma.

It is inevitable that tools get damaged while gardening. Tines can be bent from hitting stones, screws become loosened and cutting blades become blunt. Such damage can be repaired easily. Bash tines back into shape with a hammer, tighten screws with a screwdriver and sharpen blades of any kind with a sharpening stone. If you sharpen secateurs, spades and other cutting blades regularly by rubbing the cutting edge round side down with a damp sharpening stone, maintenance of the

A wheelbarrow tyre awaiting repair. (Angela Youngman)

blades will only take a few minutes. Creating sharp edges to blunt items will take longer and will require time and effort. Many garden machines can be rejuvenated very easily by oiling the moving parts or sharpening the blades.

Wheelbarrows are one of the most essential tools in a garden, but when they develop a puncture, it can be extremely frustrating as it makes the wheelbarrow much harder to push. Most wheelbarrows have pneumatic tyres filled with air. Unfortunately this also means that the wheels can be punctured very easily during normal use when weeding or pruning. Pieces of broken glass slice through tyres, and large, sharp thorns from bushes like hawthorn, blackthorn, pyracantha and roses can cause slow punctures. Eventually this results in a flat tyre running against the metal rim and making it impossible to use the wheelbarrow.

In order to deal with this type of problem, begin by checking the wheels on a regular basis. This will allow you to remove any large stones that are caught up in the tyres, as well as be able to remove thorns quickly – and with luck, before they puncture the tyre irretrievably. If the tyre has only lost air then it can be reflated by using a normal food or air pump. Leaving empty barrows turned upside down or leaning against a wall will avoid pressure on the tyres, so decreasing the risk of a flat tyre due to a lack of air.

Buying a new tyre is the usual option when faced with a completely flat one. Before doing so, it is worth checking to see if the tyre can be repaired as this is possible in many cases. Much depends on the actual type of tyre. If yours has an inner tube and a clearly identifiable valve, puncture kits can be purchased from bicycle shops in order to undertake a repair. If no inner tube is present, you will need to identify the location of the puncture and pump sealant into the hole. Soaking the tyre in a bucket of water will usually enable you to find the hole.

Keep potential problems to a minimum by undertaking a regular maintenance check around the garden. It can help to have a checklist of all the items that require periodic care – fences, buildings, greenhouses, power tools, wheelbarrows, water butts, hand tools and safety equipment including goggles and gloves. Lawnmowers, chainsaws and shredders also need regular servicing. Tick them off as they are checked and make a note of any repairs necessary.

Store tools in a dry shed or garage. Make sure they are fully cleaned before putting into storage after use. Taking basic precautions will ensure that tools last much longer, for example rubbing linseed oil into wooden handles will protect the wood from splitting. If rough areas appear, rub down with sandpaper and some oil. Leave to soak in thoroughly before reusing the tool. If garden fork prongs bend during use, knock them back into shape. If left bent, they will be less effective as a tool as the damaged area will create an area of weakness susceptible to further damage.

Tools such as secateurs, saws, scythes and loppers should be sharpened regularly. Moisten a sharpening stone with a couple of drops of linseed oil before use. When sharpening tools, push the blade forward against the stone and then to the side.

118 RECYCLING IN THE GARDEN

Oiling, cleaning and careful maintenance ensure tools stay in good condition. (Angela Youngman)

Sharpening a pruning tool.

Turn the blade over and move it against the stone, ensuring there are no rough edges. Further protection can be given by wiping blades with an oily rag after each session in the garden as plant sap and rotting vegetation will eventually cause corrosion to occur. Use a stiff brush to remove dirt from tools. The teeth of saws can be cleaned effectively using an old toothbrush. Remove grass and dirt from lawnmower blades, and ensure that the blades are tight.

Cover wooden fences and wooden garden furniture with preservative each year. This will prolong their life and overall appearance.

If you do have to buy new tools and equipment, choose carefully. Wherever possible, seek out recycled products for use in the garden as they offer greater sustainability. Shopping around is essential, but it is worth the effort. The number of products made from recycled materials, especially plastic, is increasingly rapidly. New ideas are appearing constantly, for example the gardening team at Nunnington Hall replaced the thin, single-use black plastic modules used for growing seeds with sustainable rubber plant strips sourced from Sri Lanka. Long lasting, they can be reused time and time again, rather than being a single-use system. Equally inventive are plant pots made from discarded rope and fishing nets seen at the Chelsea Flower Show in 2021.

There are many recycled hard landscaping options, which can be extremely decorative and sustainable. Reconstructed stone is a popular choice for garden features. This type of stone is manufactured from crushed recycled stone added to a concrete mix. The resultant product is moulded into a wide variety of shapes. Surface textures can vary considerably from smooth to deeply hewn so as to create different styles of stone. The blocks are then used to build seats, walls, planters, pergolas, sundials and barbeques. Other frequently used forms of recycled hard landscaping include crushed brick made from 100% recycled brick that can be used instead of gravel, and a special product called Thermalite. Solid blocks of this are made from 80% recycled, pulverized fuel ash, a coal-burning power station by-product, which is ideal for building walls.

The EcoStone product range from Deco-pak, in Halifax, West Yorkshire, includes a variety of unusual recycled aggregates. Glasglo is made from 100% recycled glass derived from cathode ray tubes used in television screens. Careful recycling enables the tubes to be broken up and turned into a translucent glass aggregate for use in water features and containers. The tumbled glass is smooth to the touch, making it safe to use even when children and animals are around. EcoStone Sea Shells is another of the company's eco-friendly recycled materials. Made from whelk and oyster shells, it is a by-product of the food industry that would otherwise end up in landfill. The shells are crushed down and made into a decorative mulch that is perfect for using in containers and garden borders. It has an added advantage in that the crushed shells act as a natural slug deterrent. EcoStone Rubber Chippings made from recycled tyres are another popular recycled landscaping material. Over 300 million tyres are sent

Patio materials and seat made out of recycled plastic. (Keder)

to landfill every year, with a tyre taking up to eighty years to decompose. Shredding tyres to resemble traditional stone chips results in a free draining, highly durable, soft and safe material that is often used in children's play areas.

Recycled plastic is becoming one of the most frequently used garden products. Turning plastic into new products uses 66% less energy than the creation of new plastic, and it can be recycled time and time again, making it a very sustainable material. Once purchased, there are no maintenance costs. It is maintenance free, and will not rot or require any preservative treatments, even when used in water or on boggy ground. In addition, the products are chemically inert and resistant to UV fading. Decking boards are more slip resistant than wood when wet, due to the slight texture and groove on the boards. Although algae might appear in damp, shaded areas, it can be removed quickly and is slow to return. This is because it cannot bed into the grain of the board. Products made from recycled plastic are very durable, strong, and hard wearing. There have even been reports that on occasions when concrete fence posts were damaged, recycled plastic fence panels experienced little evidence of any impact, even when hit by a car!

The range of products now made from recycled plastic is extensive: lumber, fencing, decking, posts, raised beds, eco paving slabs, planters, benches, shed bases, stepping stones, border edging, garden furniture, aviary panels, pet enclosures, composters, flood barriers, work pods and garden rooms. Ogel, based in Gateshead, Tyne & Wear, have devised a unique system of turning plastic waste such as car part linings, expanded polystyrene, linings from fridges and freezers into granules that are then

A recycled plastic pod created by Ogel that can be turned into a garden room or work pod.

used to construct extremely strong plastic blocks. The blocks can be combined to create custom-designed pods ideal for use as garden rooms or homeworking pods and capable of being installed and ready for use within one day. Recycled plastic has become a viable alternative to timber for most garden uses.

British Recycled Plastic, in Hebden Bridge, West Yorkshire, say that all their lumber is made from 100% recycled plastic, sourced and processed within the UK. A solid material that handles like hardwood, it comprises waste agricultural film and waste plastic from the manufacturing and automotive industries. Regular wood tools can be used on the lumber, but at a slightly lower speed to avoid any melting due to friction. The lumber can be cut across the width, but never lengthways as this will affect the structural integrity and result in warping.

When exposed to heat during the summer, recycled plastic can expand slightly and this factor should be borne in mind when measuring and planning projects. Most expansion occurs along the length of each section. Fences and border edgings should incorporate 5mm expansion gaps, thus avoiding any planks directly butting each other. This does mean that it is not suitable for creating totally watertight structures.

RECYCLING IN THE GARDEN

The quality of products made from recycled materials is very high, and can be just as good as those made from raw materials. Some materials such as aluminium and glass can be recycled indefinitely without losing their quality and bringing the concept of a circular economy into reality.

Compost made from recycled materials is a popular option among gardeners. Many local authorities take green waste and turn it into compost, which can then be purchased by consumers. Manufacturers too are creating brands of compost made from recycled waste such as hay, manure, spent mushroom compost, fruit waste, wood shavings and coir. Commercial compost production now involves an increasingly wide range of recycled by-products. Grower is a nutrient-rich material made from the by-products of a renewable energy plant fed with maize to make electricity for the National Grid. Dalefoot Composts include wool, bracken, comfrey and poo from a herd of fell ponies. Such composts are much more environmentally friendly than peat and just as effective.

Using reclaimed materials within the garden is nothing new. It is a long-established practice to take statues and columns from one garden to another, or even from one country to another. People have always taken brick and stone from disused buildings and put them to use elsewhere when building houses or paths. When strolling around villages and towns, old walls can often be seen with layers of medieval or even Roman brick and tile sourced from buildings that once stood nearby. During the seventeenth and eighteenth centuries, taking a Grand Tour of Europe was essential for many young men from aristocratic families. During those tours, they were expected to collect and send home large quantities of classical statues, columns and stone vases, which were then used to adorn their gardens. Such garden features can be admired in almost every garden attached to a historic house throughout the country. Even in the twentieth century, this type of recycling was still very much in use, as can be seen at Portmeirion in North Wales. The coastal village was designed and built by architect Sir Clough Williams-Ellis between 1925 and 1975 in the style of an Italian village. Most of the construction materials came from demolished buildings across Wales, such as a unique seventeenth-century ceiling purchased for £13 from Emral Hall, which was demolished in 1936. Williams-Ellis installed the ceiling in his new town hall. Another building, known as the Gloriette, overlooking the Piazza contains columns sourced from Hooton Hall in Flintshire. Other standalone columns from the same building now mark the corners of the Piazza. Eye-catching mermaid railings used throughout the village were originally used around the central atrium of the Liverpool Sailors' Home.

Finding suitable reclaimed materials for use in landscaping projects within modern gardens is possible. There are large quantities of such material available and it can prove far less expensive than many people would expect. Local initiatives can prove extremely useful in this respect. London-based garden designers Urban Paradise

Path made from used wooden railway sleepers.

Company have linked up with their local horticultural hub in Chalk Farm in north London. Urban Paradise inevitably find themselves with large quantities of plastic plant pots. By passing all these plant pots to the horticultural hub, the pots can be used to grow plants to be sold to bring in income, or given away to local communities for reuse. This facility is proving particularly important for people who have a low income, or are unable to easily reach garden centres, which may be located far from their homes. In addition, the pots prove extremely useful for volunteers maintaining the community public garden.

Begin by searching local car boot sales, garage sales and flea markets. Always approach such places with an open mind. You know the type of project you wish to undertake, but avoid having fixed ideas as to exactly what type of materials you wish to use. Be prepared to be creative, and think outside the box. Look at the materials and products on offer at a car boot sale, yard sale or flea market and consider how they might be used. How could they be used in your chosen project? A path could be created out of all kinds of hard landscaping, from concrete to chunks of bricks or broken stone. Wooden railway sleepers can be reused to make raised beds, edging or even paths.

In some circumstances, skips located within your local area may offer some potential material. Always check first with the householder or contractor that the desired materials are actually unwanted and that you are not trespassing on private property. It may be that the householder has placed items near a skip temporarily but plans to reuse them. Always seek the appropriate consent before taking items away.

Ebay can be a very useful source of garden items for recycling and reuse. Garden designer Sam Westcott purchased some metal gates for £200 on the site and paid a courier £60 to collect and deliver them. The gates were shot-blasted back to the raw steel for £150. A blacksmith then constructed four posts and off the shelf mild steel balls were welded to the post tops to make them more decorative at a cost of £300.

RECYCLING IN THE GARDEN

Sam points out that the completed gates had been costed at £7,000 to fabricate by three separate blacksmiths, making the recycled version a real bargain. To take another example of successful Ebay recycling, landscape architect Lucy Marshall of CW Studio in Manchester purchased a greenhouse on the site. She says:

> The greenhouse purchase was preceded by a lot of research into different brands to ensure quality and the likelihood of getting replacement parts and additional items (like staging) that would fit. Horticultural glass comes in standard sizes and replacement panels are easy to source. Some places will offer a cutting service if needed. The key thing to check is that it is Grade A toughened glass to EN12150. As with all large Ebay purchases, it's essential to see it before you buy and ask the seller lots of questions. We also bought the garage second-hand and had to get a team of friends to help dismantle it, load it onto the hired flatbed truck and then reassemble in the garden.

Reclamation, salvage or recycle yards are one of the best sources of large quantities of reclaimed bricks, metal piping and other building materials when making garden structures or undertaking hard landscaping projects such as constructing walls or patios. These yards only sell materials reclaimed from other sources such as demolished buildings. It is the ideal way to find something that will match the colour of existing brick or stone work in an older house. Salvage yards can also contain a vast array of unusual materials and waste sourced from hotels, industry and commercial buildings. Thinking outside the box is important when looking at this type of merchandise – stainless steel equipment originally intended for sterilising instruments could easily be turned into candle holders or garden features. A metal container could become a base for table when combined with a circle or square of wood or glass. At one such yard, Matthew Levesque discovered a group of glass cubes complete with metal square edges, which had originally been used as lighting fixtures within a hotel. He turned these objects into paving material for a patio by bedding the cubes in sand and backfilling with gravel, leaving the metal squares and glass edges outlined on the surface.

SalvoWeb is the ideal way of identifying the location of architectural salvage and reclamation yards. It has been active since 1991, promoting the reuse of materials from demolition, and acts as a marketplace linking companies dealing with architectural antiques, garden, decorative, salvage and reclaimed building materials. The SalvoWeb directory covers around 1,000 UK businesses, and 2,500 worldwide. There is extensive demand for reclaimed materials to use in gardens, ranging from building materials to seating and decorative items. Typical examples include Watling Reclamation, a supplier of salvage, reclaimed building materials, planters and pots, and Ronson Reclaim, offering garden ornaments and architectural stone. The range of products on offer at English Salvage includes reclaimed bricks, antique garden ornaments and architectural stone as well as the quirky and unusual such as a giant cauldron turned into a fountain and galvanised bath planters.

Community social media pages on platforms including Facebook and Next Door can also be a valuable source of information and product availability. Placing a query on such sites often results in suggestions from people within the local area who are sometimes trying to get rid of the exact type of product you are seeking.

There are a number of recycling websites available that can act as a way of sourcing materials or disposing of unwanted items such as garden furniture, tools and powered equipment that are no longer required, but other people could use. The Freecycle Network is by far the largest of these groups. Founded in America in 2003, it has since spread to over 110 countries with over 9 million members. A non-profit membership organisation, anyone can join a local group. Members have access to regular updates highlighting the availability of products, as well as being able to make requests for items required. So for example, a member seeking some planters could check the list of products on offer, as well as putting up a request for anyone with surplus planters to get in contact. The only cost involved is that of collection.

Every gardener inevitably gains more plants than originally anticipated. Plants self-seed around the garden, shrubs may send out roots that grow into new plants, and there are times when an area of garden has to be replaced, leaving existing plants homeless. All these plants can be given new homes. There may be spaces within your existing garden that can be usefully filled. Alternatively, seek new homes for those unwanted plants. Ask neighbours if they require any surplus plants, put a message on community social media pages, or sell them for charity.

Upcycling and Case Studies

Even broken or damaged items can be reused in a different format. Upcycling can make a difference and is very much in vogue. The concept is simple – a product no longer needed for one purpose, or is damaged, can be altered for a different use. A damaged, leaking rain butt could be given new life as a rabbit- and deer-proof barrier around a newly planted apple tree. Most types of damaged timber can be cut up and used to create decking, planters and wooden paths as long as they are coated with a suitable preservative. An old car can make a very warm, comfortable hen house. At Arundel Castle, old planks of wood from the drawbridge were upcycled to create a door for a boathouse on the Stew Ponds. While in Kingsbridge, Devon, Gray Patterson of ED & Dale Garden Design used both recycling and upcycling techniques to complete a garden design. Recycled materials were used to make retaining walls and fill the interiors of metal gabion cages to create garden features. Unwanted sheep troughs were upcycled to make garden ponds, with old metal lids forming simple water features.

Even children are taking the upcycling message to heart, as the example of six-year-old Abigail from Norfolk shows. She set up her own small business, Party Blooms, recycling cans into decorative pots to hold pencils and plants during the 2020 Covid-19 pandemic lockdowns. She told a local newsletter, Folk Features, what happened. 'It all started during the last lockdown. For home learning I was asked to make something out of recycled materials. Mummy and I made a pencil pot from a tin and paper napkins. I loved it so much that with Mummy's help I made an instructional video.'

Abigail continued, 'She posted it to Facebook and all her friends loved the tins. I made a few to organise my pens and pencils in my craft area and Mummy started using them as plant pots too! A few people I know got in touch with my mum and ordered some as birthday presents. That is when I decided to go into business and Party Blooms was born. Because we were in lockdown, I had more time at home. Every time we would have baked beans or spaghetti hoops, I would recycle the tin into a beautiful pot.

'I waited patiently for a chance to open my business. I wanted to have a stall outside my house – and I knew that non-essential shops were not allowed to open at that time.'

Decorative tins recycled into planters by a six-year-old girl. (Moore)

Eventually lockdown restrictions were eased, and Abigail was able to start selling pots and little plants she had grown from clippings and seeds at her garden gate. Within two days, she had sold all her pots. This gave her enough money to fulfil her dream to buy a fish tank and some fish. She still had money left over to save and invest in more materials to create more pots for sale. Her concept was simple but

RECYCLING IN THE GARDEN

effective. Each tin was covered in a mix of glue and acrylic paint, topped with a layer of kitchen towel, followed by more glue and acrylic paint to create a hard glossy surface. Decorations were added and also covered in layers of glue and acrylic.

Upcycling of this nature is not confined to domestic gardens. Even a garden centre has introduced upcycling into its trading format. Over in Australia, the Mayflower Orchids and Garden Centre in Wanneroo, near Perth, used repurposed and upcycled materials to create its business premises. The garden centre incorporates a tropical rainforest greenhouse containing a large selection of ferns, orchids and indoor plants. There is also an outdoor rainforest area, complete with native plants, succulents and fruit trees. The garden centre owners, Ray and Charlotte Maisey, are responsible for growing almost all the plants on sale. The plants are grown in recycled pots, or pots that are purchased as 'seconds quality' from a supplier. If the Maiseys had not acquired them, the pots marked 'seconds quality' would have been destroyed due to the presence of imperfections and defects.

When creating the garden centre, they were determined to recycle as much as possible. They were so successful in this aim that only 10% of the centre was constructed using new products – all of which were required in order to comply with building regulations. Old bitumen was used to create the parking area and paths, together with limestone blocks diverted from landfill. All the paving is totally recycled.

Walls and fences were constructed from old pallets and plants were hung from the sides. Gaps in the pallets were filled with planter boxes. Hearing that a neighbour was seeking to dispose of a large quantity of bricks, the Maiseys got in touch and were given the bricks free of charge. Ray Maisey says, 'We purchased a second-hand carport and transformed it into our entrance, and used recycled cool room panels and doors to close it off.'

Metal door frames and pool fencing were repurposed to construct pergolas. A customer donated palm fronds to cover the top of the greenhouse pergola. The upturned frame of a glass table became a display unit for hanging plants.

'We have used recycled materials everywhere. It's a work in progress, but it's creating a nice atmosphere for people to walk around. People sometimes just sit down on the bench and sit there for ages getting a feel for the place,' commented Charlotte Maisey.

With a little bit of thought and imagination it is clearly possible to create quite large projects such as greenhouses, sheds and planters out of material that might otherwise have been consigned to landfill. Among the many examples that exist within the industry are:

Apple storage chest

Apples need to be stored with care in the autumn. Air needs to circulate around them, yet they have to stay dry and away from light. An unwanted chest of drawers into an attractive apple storage chest. Having thoroughly cleaned the furniture, he prepared the wood surface and repainted it white. A cross pattern of circles were drilled into the sides of each drawer to create a constant flow of air. Each circle was approximately 2cm (1in) in diameter. The circles were drawn onto the wood, before being cut out with a saw drill bit. Apples could be stored safely in the well-aired shade provided by the drawers. Access was easy – just pull open the drawer.

Dead hedges

Tree branches, woody pruning material, even dying Christmas trees have been given a new lease of life by creating a very different type of hedging. Eco-friendly dead hedges are a type of man-made structure that pre-date the Domesday Book. There is even some evidence for the presence of dead hedges during the Bronze Age, when their function would have been to act as a barrier to protect livestock. In recent years, the concept of dead hedges has undergone a major revival and they are increasingly common in corners of gardens sheltering compost bins, sheds and vegetable areas, in fields or as windbreaks around wildlife corners. They can also be sited in places where a normal hedge might not grow, such as in deep shade, or on poor soil. Dead hedges offer significant environmental advantages since they provide shelter for small creatures and birds, nesting sites and a safe hibernation spot for hedgehogs. In addition, a dead hedge acts as a natural habitat and corridor for wildlife within an area.

Dead hedge. (Angela Youngman)

Norfolk gardener Christina Wakeford explains how she created her dead hedge as a shelter for a wildlife area at the edge of her garden.

Drive in suitable posts two or three feet apart in a double row. Make it any length – straight or curved; mine is about 3ft wide and 5ft high. Then begin to fill the row with wood prunings, laying them horizontally and pushing them down as far as you can go. As they accumulate, the layers look attractive in a rugged kind of way. For a neater look, you could weave some pliable prunings (coloured dogwood or willow) through the posts along the length. The woody material decays extremely slowly and may be continually topped up; I'm still adding to the one I started in 2010. It is so easy and environmentally friendly and can be adapted to any shape/size/design suitable for your garden.

Other examples in her area include dead hedges formed from the waste material generated by pruning fast-growing poplar trees, hay and beech hedges.

Wooden drawers into planters

Adapting a chest of drawers or individual drawers to make planters is extremely common. They can look very stylish, and offer flexibility in the way they are presented. Garden designer Simon Orchard of UK-based Orchard Design discovered

Individual wooden drawers turned into planters.

an unwanted chest of drawers in the street, took it home and upcycled it to make a very attractive garden feature. Having cleaned it up, he set out to make the drawers watertight. Each drawer was pulled out a little way and the inside covered with landscape-quality polythene. After adding compost, it was planted up with pretty trailing annuals such as lobelia.

Fencing and seating

In his book *Garden Eco-chic*, Matthew Levesque suggests creating fences and screens out of doors, fibreglass rods, woven sections of pallet strapping, old surfboards, salvaged steel wire ties linked together to create a textural screen when hung in horizontal rows, and even a group of old shovels planted blade upwards in rows. He even creates a pond wall surround out of stucco-covered polystyrene possessing the appearance of stone. For seating he recommends combinations of aluminium tubes, wooden boxes, pallets, milk crates, shipping crates and chunks of stone to provide legs.

Herb planter made from metal shelves and ironing boards

The shelves came from a damaged mini greenhouse. The plastic covering on the shelves was peeling off and the metal interior had become rusted. The ironing boards were rusty and corroded. All the rusted surfaces were scrubbed clean with sandpaper to create a smooth surface and then painted with rust converter paint. This turned the rusty areas into a solid mass that would undergo no further deterioration. When the paint was fully dry and checked to ensure that there were no further rusty areas that had been missed in the original cleaning, the new planter was assembled.

One ironing board was stretched out so that it formed an X on the ground. The second ironing board was placed across the bottom of the legs, and the two metal shelves formed the remaining sides. The sides were firmly tied together. Four decorative planting areas were created that could be used for herbs, vegetables or annuals.

Plastic bottle greenhouse

Greenhouses can be expensive to buy, especially for schools operating on a limited budget. Learning how to grow food is part of the curriculum, with many schools having small growing areas on site or in nearby allotments. Watching plants grow and develop from seeds is an essential gardening task. By creating greenhouses out of plastic bottles, schools are able to combine gardening with teaching about design, recycling and sustainability as well as providing a much-needed facility at very little cost.

Inexpensive and simple to create, bottle greenhouses have become a feature of most school gardens. The Royal Horticultural Society offers a suggested version on its schools website. Over the years, schools and groups have created many different

A simple greenhouse made out of plastic bottles.

variations and no two greenhouses are ever the same. The amount of bottles required depends entirely on the size of the planned greenhouse; one 8ft × 6ft might require up to 1,400 empty 2 litre bottles.

At its most basic, a plastic bottle greenhouse requires a basic framework, which can be purpose-built out of recycled wood by parents, or use a repurposed old framework. A line of holes is drilled along the top and bottom frames, large enough to hold a series of long bamboo canes. Having collected a large quantity of bottles, the base of each one is removed so that it can be slid down a cane. More bottles are added until the top is reached. The interlocking bottles form a solid column. The column is then inserted into the top hole of the framework and fixed with a cable tie. More columns are added until the sides are completely filled in. At the front and back of the greenhouse, the columns are graduated to allow for a slanting roof. A door is added, sometimes made out of an old window or more bottles. Typical roof materials include more bottle columns or large pieces of polycarbonate. Packed closely together, the bottles provide solidity, warmth as well as double glazing, ensuring a higher temperature inside than out and thus encouraging seed growth. When the bottles deteriorate or are broken by accident, they can be easily replaced since all that is required is to unhook the line, remove the bottle and replace it with a new one. A layer of bubble wrap across the interior walls adds extra insulation.

Pallets

These are popular items for upcycling around the garden. The best pallets to use are ones that contain the IPPC or EPAL logo plus the letters HT – this ensures the wood is free from any toxic pesticides and has been heat-treated for long-term use.

One of the simplest projects is to turn four pallets into a compost bin. Stand the pallets on edge to create four sides. Fasten them together with strong garden wire. The natural gaps in the pallet provide a constant supply of air to the decomposing compost.

Other upcycling projects have included creating pallet sofas by combining two or three pallets for the base and one for the back rest. A stack of two or more pallets can create a simple rustic table. Laid on the ground at a slight angle, ensuring that water can drain away, a pallet can make an unusual setting for rockery plants. Place a waterproof membrane underneath, add compost and plant low-growing rockery perennials between the slats.

Placing pallets on end against a wall creates vertical planters. Painted up in bright colours, they can make an attractive garden feature. Simon Orchard of Orchard Design explained how to undertake such a task. A claw hammer is used to remove every second board from the front of the pallet in order to provide room for growing plants. Depending on the exact location of the boards, some of the original ones may have to be nailed in place to ensure a matching location front and back. This ensures that the plant pockets can be fully attached. Cover the back, sides and base of the

Wooden pallets painted in varying colours make an attractive display area for flower pots.

pallet with a weed mat. Use Blu Tack to hold it in place while you check the fit. Once happy with the result, staple the mat in place and remove the Blu Tack. Trim off any excess so that the edges are neat and tidy. Measure and cut weed mat pockets so that they fit inside the vertical supports, then add some drainage holes, fastening them in place before filling with soil and plants.

Pallet collars are equally useful and can often be found on Ebay. Each collar comprises the four exterior sides of a pallet, joined together with metal fastenings at the corners. Successive layers of collars can be placed on top of each other to create compost bins. Alternatively, they can be used singly to form a raised bed or planter.

The creation of junk sculptures is a steadily increasing trend. All it involves is some imagination and a selection of unwanted materials. It can result in some very eye-catching and decorative items, such as a giant ladybird made out of a red safety hat and wire hangers to make legs. Combining a selection of different-sized empty plant pots with some wire and paint can create very popular plant pot people. This makes an interesting project for children, and has even resulted in the creation of a temporary sculptural garden trail around the town of Settle in North Yorkshire.

A turtle made out of garden junk.

On a much larger scale a few years ago, a GoGoDragons Wild At Art sculpture trail in Norwich involved a salvage yard commissioning an artist to take part in the trail, creating a massive dragon using just the materials available within the yard. In 2021, a GoGoDiscover Wild at Art Junkasaurus Rex sculpture was collaged with pieces of waste material.

The vintage look is equally popular with people wanting to incorporate vintage styles and designs in their homes and gardens. Upcycling items to create the impression of a patina of age forms part of that vintage style. Although iron and other metal objects will eventually possess a rusty appearance over time, it is possible to speed up the process by using a mix of sandpaper, pool cleaner, hydrochloric acid or acetic acid. To do this, the object has to be placed in a solution of pool cleaner

GoGoDiscover Junkasaurus T-Rex covered in waste materials.

(¼ cleaner to ¾ of water) and a solution of hydrochloric acid or acetic acid (5–10% acid to 90–95% water). After the allotted time has been reached, the object should be removed from the solution and rinsed thoroughly with fresh water. Roughen the metal with sandpaper as this will encourage rain water to penetrate the metal. Place in the garden and leave to the elements. While it will take time for the object to become really rusty, some changes will begin to appear after the first rain shower.

Case study: Upcycling at Antiques by Design

Guy Chenevix-Trench is an expert at upcycling garden items. He is an artisan on the BBC programme *Money for Nothing*, as well as running Antiques by Design. He says, 'When I look at something even as simple as an oil drum, I get thousands of ideas on how to use them and what can be done with it.'

The first task is to clean the item to remove dirt and residues such as oil using wire wool. Many marks are retained deliberately as they form part of the product's history. For example, paint chips on an oil drum fade with time and can look very attractive. Whatever the item, whether wood or metal, once the initial cleaning is complete, a coat of beeswax is used to cover the entire surface. Guy leaves the beeswax on for five minutes, before buffing it up to create a shine. This is sometimes repeated up to two more times, depending on the extent to which the age of the item is being accentuated.

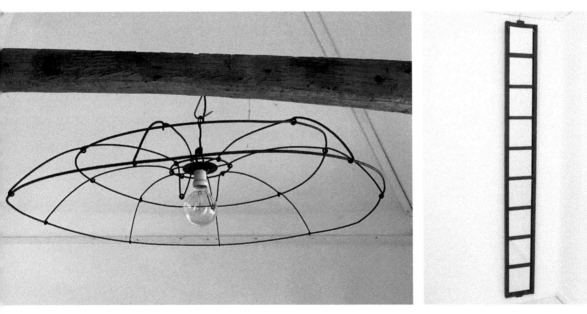

Above left: *Victorian rose support becomes a dramatic light fitting after conversion by Antiques by Design.*

Above right: *A traditional barn ladder turned into an eye-catching mirror. (Antiques by Design)*

His approach has led to the creation of some unusual, unique items. Visiting an auction, he acquired a large barn ladder, which he turned into a series of mirrors. After cleaning the ladder with beeswax, sheets of individually cut glass were placed between each of the steps on the ladder. These were fastened securely in place, and a fixing added so that the ladder could be hung on a wall.

On another occasion, an oil drum was cleaned and polished before being cut down to make a table. Mirror glass was added to the outer ring, creating a very fashionable mirror that could be used inside or outside the house. Similar mirrors were created from the tops of galvanised water storage tanks.

Garden tools are among the most popular items Guy uses to upcycle into lighting fixtures. Family members often approach him to upcycle a cherished fork or trowel in memory of a beloved husband or wife. Guy states,

> The tools can be permanently fixed in place, or we can create a U-shaped cradle into which the tool can be lodged securely against the central pole, so that the tool can be taken and used when required. I have made many lights of this kind. I brought an antique pair of Victorian garden shears and cleaned them up with wire wool, then beeswax. The blades were left blunt. I added a two-tier granite base using offcuts from a specialist company – a 20mm bottom, followed by a 10mm piece on top. The shears were fixed to a pole, and an electrician added wiring, a bayonet fitting and undertook a PAT test. We then added a shade to complete the light fitting.

A similar technique has been used to create lights from full-size garden forks and spades, flower pot lamps, a cloche, an antique rose support and a copper French garden sprayer. One of the most unusual lights was created by accident. Guy purchased a metal watering can at a sale because he liked it, and used it in his garden for over ten years. It eventually developed a leak and he was told by his wife to either repair it or do something with it. He decided to upcycle the watering can

Guy Chenevix-Trench upcycled a metal watering can into a light fitting to reveal a Nazi swastika on the base, making it extremely rare. (Antiques by Design)

into a light fitting. On cleaning the can, he discovered a Nazi swastika on the base, marking it as a rare object that was manufactured in Germany during the 1930s.

Case study: Sylvan Studio

Asked to design a new garden around a new barn conversion, Christine Whatley of Sylvan Studio in Wiltshire was able to upcycle numerous agricultural items left in the vicinity from its previous life such as an old galvanised iron water trough, which was turned into a decorative planter. She says,

> There was lots of stuff lying around such as a granite mantel left in a field, and a grate I discovered when rummaging among the rubbish. The clients asked me to find a use for them. They were incorporated into the design at the initial stage, and in fact led the design to a certain extent. The grate was built into the paving by setting it onto a brick lip so that it finished flush with the surrounding paving. A gravel infill was used within the grate.

Upcycling the granite mantel required some imagination. Christine used the surrounding landscape as inspiration to turn it into a very attractive bench. A metal frame was made to fix the mantel in place, and wooden slats were placed on top. The slats were deliberately cut to reflect the shape of the skyline behind the house, based on a tracing from a photograph. The base of the mantel was set slightly below the surrounding paving level in order to achieve a good sitting height.

Reflecting the skyline using a granite mantle and wooden slats to create a seat. (Christine Whatley Sylvan Studio)

Case study: Refugee and industrial waste-inspired gardens

In 2018, designer Tom Massey created an innovative RHS Chelsea Flower Show garden inspired by a visit to Syrian refugees living in the Domiz 1 camp in northern Iraq. The Lemon Tree Trust garden was designed using input from refugees, who had created unexpected horticultural beauty within the confines of the camp. Using only recycled materials in ingenious ways to develop garden spaces that provided relaxing, productive gardens, camp residents used gardening to restore a sense of normality and peace to their broken lives.

For the show garden, Tom says he used 'ingenious vertical planting, inspired by refugees' use of everyday objects, providing ideas for planting in limited space. Brutal harsh materials, such as concrete and steel, widely available in the camps, were elevated with techniques such as polishing, casting and crafting into patterns and intricate Islamic-inspired designs. Cooling and calming water flowed through the space, collected in channels and pools, recycled and pumped back through the brimming central Islamic-inspired fountain, representing the importance of grey water reuse and the many makeshift fountains refugees have built in their own gardens in Domiz camp.'

As another example of the potential of recycling techniques within garden design, Tom created a garden for a BBC2 programme, *Your Garden Made Perfect*, in which a range of upcycled materials were used, along with waste building material as a low-fertility substrate for planting. The design had to meet very specific requirements as the owner of the garden was a wheelchair user; all the hard surfaces had to be easily negotiable and step-free with ramps of no less than a 1:12 gradient. Low-cost industrial construction materials such as huge concrete pipes for planters, galvanised ducting, tin roofing, scaffold boards and 'soil' made from crushed concrete, shingle and sand demonstrated how to repurpose or upcycle materials designed for other uses. The materials chosen deliberately referenced the wider dockside/industrial landscape of Littlehampton, West Sussex, and nearby shingle sandy beaches. The planting was inspired by Mediterranean and brownfield sites that possess a similarly hostile environment characterised by harsh, arid conditions and very infertile, poor soils yet are havens for wildlife especially insects, providing habitats and food sources.

Tom states,

The garden replicates a 'disturbed' environment, planting in waste materials, which are nutrient poor. The key elements are relatively simple: varied topography and low-nutrient soils. It makes them relatively maintenance free as most weeds won't like the substrate and prefer fertile soils. The 'wild' look requires little intensive pruning and maintenance. It provides an accessible, unusual, wildlife-friendly haven for the owners, which they can all enjoy together with intimate and secluded spaces to satisfy all of their needs.

Case study: Melanie's urban garden

On moving into a new house, there was no budget for landscaping the garden so garden designer Melanie Debenham chose to use as much reclaimed material as possible. There were already concrete slabs present, which the builders moved to the far end of the garden to create a patio. Discovering gravel from a local garden clearance being thrown out, Melanie reused it in the garden. Over the next few years, the garden was developed in stages, beginning by building raised beds at the rear and a seating area, which took into account the presence of a mature oak tree. Old breeze blocks together with pieces of iroko wood left over from Melanie's previous house were used to make a seat. Joists removed from the house during building work became a base for raised decking, together with some screens left over from landscaping around her son's school.

Although being advised to simply replace the shed, Melanie and her husband dismantled it and moved it to a better spot in the garden, where it was given a new roof. Galvanised water tanks were upcycled into a planter and water feature. A coating of lead paint was added to the water feature to reduce the risk of corrosion. A supply of bricks for edging borders was obtained from a builder working on a nearby house. Discovering that the builders were throwing out a quantity of glazed

Melanie Debenham's urban garden created using recycled materials. (Melanie Debenham)

A galvanised tank turned into a planter. (Melanie Debenham)

bricks in a pretty brown sienna colour, some of which were striped in white, Melanie spoke to the builders and offered to take them away. All she had to do was collect and move them. Even plants were reused – two huge buddleja that had outgrown a friend's garden was transferred to its new home in Melanie's garden, instantly helping to bring it to life.

An unusual feature of the garden development was the use of a washing machine rotor drum as a fire pit. Melanie says, 'We had seen this done online, and decided to try. We got one from a salvage centre and placed it so that it is at table height. There are some downsides, it doesn't work as brilliantly as we hoped because you can only use small logs in it and it creates a lot of smoke.'

Looking back on the garden renovation, Melanie says that the hardest task was building the deck, putting the screens up and concreting in the posts as these were tasks she had never undertaken before. She says, 'You need to be as thoughtful and creative when reusing and upcycling. It is important to be realistic about what's possible, as it can be heavy and hard work.'

Case study: Caspian's cat shelf

Using some leftover wood, Marie Shallcross of Plews Garden Design in Beckenham, Kent, created the ideal solution to cater for the fence-jumping activities of her beloved

Caspian on cat shelf showing its construction and vegetation growing around it. (Marie Shawcross)

cat, Caspian. Caspian enjoyed sitting on top of a tall fence post that provided a high viewpoint covering the garden. Caspian frequently jumped from the top of the fence onto a granite patio and although he was only two years old, Marie was worried about long-term injuries from the impact his legs experienced on landing. Once or twice he had landed awkwardly and sprained his leg. Some oak decking left over from a garden design and landscaping project together with some shelf brackets hanging around in the garage provided the answer.

Honeysuckle adorning the fence was pruned back and Marie used the wood to create a cat shelf situated part of the way up the fence. The shelf was carefully placed at the exact spot where Caspian reached down when feeling less inclined to jump from the top of the post. Getting the right height was important in ensuring that it was comfortable for him to use. The new cat shelf was instantly popular. As it caught the morning sun, Caspian used it as a seating spot after breakfast, while his litter sister Mulan also used it from time to time.

Marie is a keen upcycler, and has used unwanted garden products in new ways on many occasions. An old folding outdoor rabbit run was recycled as a cold frame

Above: *Caspian's cat shelf just after it had been put up. (Marie Shawcross)*

Below: *Hens living in a refurbished hen house. (Marie Shawcross)*

on her allotment. Added warmth was provided by fleece tacked to the interior sides and removable top. To reduce the risk of foxes playing with the horticultural fleece as well as providing slug and snail prevention, citrus peel was placed around the cold frame.

Rotting wood on a hen ark needed to be refurbished. Pieces of oak decking, together with other timber left over from other projects, was used on the ark itself. An old wooden climbing frame/treehouse was dismantled and used to form two runs at each end of the ark. Other recycled materials used on the project included paling fencing and wire from puppy-proofing the hedge boundaries. The wire was used on the edges of the run floor and below ground to prevent digging. The only new items required were narrow wire mesh for the walls and roof, nails and door hinges.

Case Study: Gardening on a shoestring

London-based garden designer Lisa Toth had long wanted to explore the concept of garden design on a shoestring, pointing out that,

> People don't realise how much work is involved in creating a garden using recycled items. It takes time to find the materials, especially if you have to scavenge from skips and neighbours. Those materials may cost very little, but it takes time and labour to do the actual construction work. If you go to a salvage or reclamation centre, you do have to buy the products. A recycled garden is not an instant or necessarily cheap option.

The start of the Covid-19 pandemic and the resultant lockdown provided the ideal opportunity for her to test out recycling concepts. Lisa set out to redesign elements within her garden, building a patio and pergola from recycled bricks, pavers, tiles and wood. She had been collecting potential material for some time, storing it in the garden, sheds and putting planks in the hallway. First creating a sketch on paper, she then went out into the garden to explore the options in practice. The key aim was to create a seating area that would allow them to enjoy the evening sun. Surplus bricks were used to mark out a circle on the ground until they were satisfied with the result. The first task was to remove a dead plum tree, and dig out the foundations by removing the existing lawn. Excavated soil was placed to one side and left to rot down for reuse on the vegetable beds. Any stones and bricks discovered during excavation were added to the hard core for the foundation.

The design called for a brick edging to be placed around the seating area. Although they already had a collection of bricks to reuse, there was not enough to complete the project. Lisa placed a message on Facebook asking friends and neighbours if they had any unwanted materials, especially bricks. A friend responded immediately, offering a load of bricks; all they had to do was collect them. Having undertaken a bricklaying course earlier in her career, Lisa was able to use the skills she had learned to install the bricks in place.

For the seating area itself, the intention was to create a mosaic out of a miscellaneous collection of bricks, tiles and pavers. They already had some Indian sandstone pavers and were given more by a neighbour who had replaced his patio. The mosaic outline was laid out on the lawn and altered until it suited their requirements. A layer of mortar provided the base to set the various pieces in place replicating the chosen design. A key part of the design was reusing several decorative tiles that had formerly been part of a fire surround. The problem was that they were thinner than the sandstone pavers. To overcome this situation, Lisa created casts of the paver in a wooden mould, covering the base with a liner. Each cast was filled with mortar and a tile placed within the mortar. Once the material had set, the mould was removed and the new paver (330mm square and 45mm thick) could be inserted into the design.

A pergola was designed to complement the new patio. Almost all the timbers used in the construction were available on site for immediate reuse, although some chunky uprights did have to be purchased. The uprights were set at varying levels, covered by wooden slats to create the impression of a wave. Scaffold boards were used to build a seat. A stepping stone pathway made from recycled sandstone was created. To provide extra privacy from neighbouring houses, an old gate found in a skip was reused as a frame for climbing plants.

Designing and building the patio and pergola, as well as sourcing the remaining materials, took one month to complete.

Case study: Warner Brothers Wonderworks nursery garden

Tim Jennings of Everchanging Garden Design in Hertfordshire was commissioned to create an outside space for a new nursery known as The Wonderworks, to be used by the children of staff and actors working on Warner Brothers film productions at the company's site in Leavesden. Upcycling was regarded as an essential part of the overall design and involved the transformation of a tarmac road. Having removed the road, Everchanging Garden Design set about transforming the site. Some of the road construction materials were used as hardcore for the paths.

Tim's design was based on wild meadows and woodland planting, with winding pathways leading between sections. It was decided to use oak trees that had been blown down naturally and would otherwise have been used for firewood. A mobile sawmill was brought to the site to cut the trunks and branches to match their requirements. All the architectural landscaping features including upright stakes, seats, benches, climbing frames and fences were created from these trees. Any remnants left over

Above: *Log seating area, fencing and garden features made out of recycling wood at The Wonderworks nursery, Warner Bros Studios. (Everchanging Garden Design)*

Below: *Wine boxes turned into tiered planters at The Wonderworks nursery. (Everchanging Garden Design)*

were immediately reduced to chippings for use on the paths. Old scaffold boards were re-sanded and used on the fences. Hazel hurdles were handmade on site, as well as installations for use as little dens.

Old wine boxes were repurposed as planters full of herbs with spaces to interplant vegetables for all seasons. A willow ring frames the entire garden view. Designing and building the project took around four months to complete and resulted in a very pleasing environment. It created a green space where previously there had been nothing. Wildlife began appearing within minutes of the garden being completed, as bumblebees and butterflies made their appearance, quickly followed by foxes.

Case study: Breaker's yard

Sutton House, Hackney, is a Tudor house in the care of the National Trust. The adjacent garden area has a very different ambiance. From 1920 to 1990 it was turned into a breaker's yard filled with scrap metal from vehicles of all kinds. After the yard was closed and the area cleared, the National Trust sought opinions from the local community as to the type of garden they would like to be created within the available space. There was a universal desire to reflect the yard's twentieth-century history. The result is The Breakers Yard, designed by Daniel Lobb. He states: 'The site was very overgrown and unkempt. The ground was contaminated with waste engine oil and heavy metals to a depth of up to 7m. An impermeable membrane had to be placed across the entire site, and all planting had to be in containers.'

A herringbone brick path complete with chevron designs winds its way through the garden, linking both the Tudor past and the twentieth-century history. A matching Chevron design can be seen on the massive planters created out of tyres, which house plants such as yew, dwarf fan palm and box. Dominating the breaker's yard are two upcycled vehicles. In the centre of the yard is a double-decker caravan loosely modelled on the style of a boat. On entering the caravan, visitors discover a stately home-style interior complete with replica Adam staircase and a chandelier. The other vehicle is a 1980s coach, which was purchased on the internet. It had been used as a royal staff transport vehicle, transporting people from Horse Guards Parade to Buckingham Palace. This has been turned into a greenhouse. Elsewhere on site, a repurposed tyre trailer has been turned into a raised bed as it is the ideal height for disabled gardeners. The trailer provides a good depth of soil and is used to plant herbs such as oregano and thyme as well as colourful daisies. A tool chest from the back of a pick-up truck has become a bug home, while other containers provide accommodation for an orchard avenue comprising dwarf apple trees native to south-east England. A sedum roof helps capture rain water, which drains into a large barrel from which children can pump it out into a rill running through the garden.

Case study: Beeview Farm, Wales

Matthew Watkinson has become the consummate upcycler. He and his wife Charis moved to live off grid, on a zero-carbon farm in Pembrokeshire, Wales. Their home comprises a stripped out horse lorry, a couple of caravans and two big flatbed trailers that were destined for scrap. There are no mains services such as electricity, water or sewage. Finding ways to recycle and upcycle products has made their project possible. Matthew states,

> I started experimenting because I wanted to see what was practical and possible, free and cheap. I can see beauty in waste, turning it into something new with a world of possibilities. I started with an old radiator, painted it black and linked it to a bucket to heat water by the sun. Now everything is regarded as potential upcycling material.

Large 200 litre plastic containers that formerly held chemicals for use on farms have been cut up to create overlapping tiles for cladding and insulation. The final result resembles dragon scales.

Tyres, wood, greenhouse recycling at Beeview Farm, Pembrokeshire. (Matthew Watkinson)

A recycled greenhouse complete with metal gabions at Beeview Farm. (Matthew Watkinson)

IBC tank and gabion tank cages are one of Matthew's favourite recycling items. These second-hand tanks are widely obtainable as they can be purchased cheaply from farmers, or even obtained on sites such as Gumtree and Ebay. The cost is low as long as they can be collected in person. One of Matthew's first projects was turning a cage into a woodstore. This was undertaken by turning the cage on its side and covering the top with polythene held down with some timber. Wind is allowed to pass through the cage, thus drying the wood, but it is protected from the rain. Another cage was turned into a solar kiln with kindling placed inside and left to dry by the heat of the sun.

Three linked IBC tanks were used to create a reed bed system, which cleans grey water from showers and sinks. The water moves by gravity between the beds. Each tank is filled with gravel together with bulrush and yellow iris. The waste water filters over the gravel, with nutrients being left behind and taken up by the plants. The deep green water and lush green growth in the first tank reflects the amount of nutrients being taken from the grey water. Each tank onwards is steadily more clear and growth less lush. The final water is used around the fruit trees. On average, the reed bed system copes with between 100 and 150 litres of grey water per day.

Yet another IBC tank has been turned into a biodigester, dealing with food waste and weeds such as nettle and dock leaves. Some cow manure is added to the system. As the material decomposes, it gives off methane gas, which is used as a fossil-free cooking fuel, while the remainder becomes a natural fertiliser.

Composting is regarded as extremely important. As the property is off the grid, Matthew and Charis have responsibility for all waste and energy use. A compost toilet deals with human waste, and the contents are mixed with crushed, dried bracken collected from the adjacent fields, before being used as a soil conditioner. The bracken is collected and placed in large sacks, and left to dry and break down into tiny pieces. Bracken is also collected for bedding for the chickens. The resultant bracken–chicken manure mix is placed in tractor tyre bins to break down for compost.

More bracken, brambles, nettles and weeds are harvested to create compost piles in the chicken paddocks. A compost pile is left to rot down for three months, and then the chickens are moved to the paddock where they spread out the compost while hunting for earthworms and other insects. During this process, they add their own

A car greenhouse and planter in situ at Beeview Farm. (Matthew Watkinson)

manure to the mix. In due course, the chickens are moved onto the next paddock and a new compost pile. Green manure is grown in the first paddock before becoming the site of another compost pile. There is a steady rotation of chickens and paddocks over the year. Matthew sees this as being a way of working with the environment, turning even brambles into something useful. Everything, even weeds, becomes a valuable resource.

Old tractor tyres have been used to create long-lasting, hard-wearing paths that are easy to maintain. To create these paths, Matthew removes the central rim, cuts through the side of the tyre and then allows it to roll out as a long paving material. Approximately thirty tyres were used to form the entire path around the farm. The only drawback is that it can get a little slippery in wintertime, but otherwise the tyre path works extremely well.

An equally innovative approach was taken towards constructing greenhouses. One greenhouse was created between the house and the stone wall of a field. With two walls providing warmth, the area just needed a roof. To form this roof, Matthew collected a group of scrap windows of varying sizes and types and fitted them together like a jigsaw puzzle.

The second greenhouse was made from an old hatchback car. The seats were removed and the floor was flattened to provide a growing area for trays of young plants. It is quite spacious, with room for over fifty plants. Windows can be opened when necessary to provide ventilation. It is mouse, slug and snail proof. The car bonnet was removed and the area filled in with compost, enabling its use as a planter. This car greenhouse is most effective as a spring growing area as it acts like a big cold frame; it can get too hot in summer and does not provide sufficient height for tall plants like tomatoes.

Conclusion

Recycling, reusing and upcycling has become a crucial factor among modern gardeners, affecting everyone from landscape designers and historic houses to householders and allotment holders. In a world where climate change and sustainability is increasingly important, gardeners can make a major difference. All it takes is a little imagination and time.

Recycled materials can provide extremely interesting, eye-catching design features. Even the simplest task, cutting down a tree, can turn an object no longer needed for its original purpose into an attractive feature. In one rural village, the householders were faced with having to remove a large tree at the front of their garden. They decided to leave a large part of the trunk in place, turning it into a very attractive feature by carving a little door and windows. It became an enchanted home for fairies and elves, making every passer-by smile. The addition of fairy lights at Christmas time added to the enchantment.

Recycling, reusing, upcycling are ways in which everyone can contribute to dealing with the climate crisis, while helping wildlife and the environment around us. Remember the children's nursery rhyme: 'For want of a nail, the shoe was lost', resulting in an ever greater series of consequences culminating in 'for want of a battle, the kingdom was lost, All for the want of a horseshoe nail'. The rhyme emphasises how seemingly simple acts can have unforeseen, long-term consequences. The same is true when it comes to recycling, reusing and upcycling. One action can have major implications since if everyone adopted this practice, it could have a massive effect. Setting an example, showing what can be done within your own garden or allotment, is a step in the right direction as it offers the potential to ultimately be multiplied many times. Who knows who might be influenced by what you have done and follows your example? Or what ideas or actions it could lead to? A simple action can make a difference.

Functional, innovative, creative and extremely eco-friendly, recycling provides untapped opportunities for every gardener and saves money.

An imaginative elf house made out of an old tree stump. (Karis Youngman)

Resources

The Balcony Gardener, Isabelle Palmer, CICO books, 2012.*Garden Eco-Chic*, Mathew
   Levesque, Timber Press, 2010.
Green Roofs, A guide to their design and installation, Angela Youngman,
The Crowood Press, 2011.
Grow Your Food for Free, Dave Hamilton, Green Books, 2011.
Reduce Reuse Recycle, An Easy Household Guide, Nicky Stott, Green Books 2006.
The Victorian Kitchen Garden, Jennifer Davies, BBC Books, 1987.
The Wartime Kitchen Garden, Jennifer Davies, BBC Books, 1993.

Antiques by design, www.antiquesbydesign.co.uk
Centre for Alternative Technology, www.cat.org.uk
Freecycle UK, www.freecycle.org
Garden Organic, www.gardenorganic.org.uk
Glacier Gardens, www.glaciergardens.com
Plantlife, www.plantlife.org.uk
Royal Horticultural Society, www.rhs.org.uk
SalvoWEB, www.salvoweb.com

Garden designers who helped with this book:

   www.cwstudio.co.uk
   www.simonsmithgld.co.uk
   www.simonorchardgardens.com
   www.thelovelygarden.co.uk
   www.samwestcottgardendesign.co.uk
   www.sylvanstudio.co.uk
   www.plewsgardendesign.co.uk
   www.tommassey.co.uk
   Lisa Toth – Instagram, @gardendesignmatters, www.gardendesignmatters.co.uk

Index

A

Apple corers, 71
Aluminium containers, 71
Antiques by Design, 94, 138, 139, 140
Arundel Castle, 27, 28
Ash, 71
Autumn leaves, 72

B

Baby Baths, 73
Bags, 73
Balls, 73
Bamboo, 73
Barrels, 73
Beeview Farm, 86, 150, 151, 152, 153
Biddulph Grange, 21, 22, 23
Bikes, 74
Biodegradable packaging, 74
Blankets, 74
Bokashi, 56
Bras, 74
Bricks, 75
Broken crockery, 75
Brushwood, 75
Bubble wrap, 75
Buckets, 76
Bug hotels, 12
Building rubble, 76, 147

C

Cable reels, 76
Cans, 76, 104
Cardboard, 76
Cardboard tubes, 77

Car parts, 77, 149, 152, 153
Carpets, 78
Cartons, 78
Cart wheels, 78
Carol Whitehead, 114
Chelsea Flower Show, 68, 69, 70, 141
Chest freezers, 78
Chicken wire, 78
Chimney Pots, 78
Chocolate box trays, 79
Christmas trees, 79
Cobbles, 79
Coffee filter papers, 79
Coffee grounds, 79
Colanders, 79
Comfrey, 79
Compact discs, 80
Compost, 18, 48, 49, 50, 51, 52, 53, 54, 55, 56, 123, 152
Compost bins, 81
Concrete, 81
Containers, 81, 150
Cooking oil, 81
Corks, 81
Corrugated cardboard, 81
Corrugated metal, 81
Corrugated plastic, 82
Cutlery, 82
Cycle helmets, 82

D

Daniel Lobb, 149, 150
Dead hedges, 130

Dishes, 83
Dustbins, 83

E
Egg boxes, 83
Egg shells, 83
Everchanging Garden Design,
    147, 148,

F
Fabrics, 84
Feathers, 84
Fencing, 84, 132
Flour shakers, 85
Foil, 85
Freecycle, 126
Fruit punnets, 85

G
Gabions, 85, 151
Glacier Gardens, 24, 25
Glass jars, 86
Green manures, 12
Gray Patterson, 127
Grey Water, 39, 40, 41
Ground Source heating, 46
Growbags, 87
Guttering, 87

H
Hats, 87
Hot beds, 48
Hollow plant stems, 88
Hot water bottles, 88

I
Ice cream containers, 88
Ickworth, 23, 24
Ironing boards, 88, 132

J
Jiffy bags, 88
Junk sculptures, 135, 136, 137

K
Kitchen paper, 88
Kusan Doi, 37, 38

L
Ladders, 89, 139
Lisa Toth, 146, 147
Logs, 12, 89, 90
Louvre panels, 90
Lucy Marshall, 125

M
Manure, 90
Marie Shawcross, 83, 84, 143, 144,
    145, 146
Mayflower Orchids & Garden
    Centre, 129
Melanie Debenham, 142, 143
Milk cartons, 91
Mortar, 92
Moss, 92

N
Net curtains, 92
Nettles, 92
Newspaper, 93

O
Oil drums, 94
Organic Garden Company, 116
Orchard Design, 134

P
Paint brushes, 95
Pallets, 95, 134
Party Blooms, 127, 128, 129
Paper, 96
Peat, 7
Pet bedding, 96
Plastic, 96, 97, 98, 121, 122, 124, 132,
    133, 150
Polystyrene, 98
Pots, 99, 104

R

Railway sleepers, 100
Rainwater recycling systems, 32, 33, 34, 35, 36, 37, 38
Rajul Shah, 25, 26, 112
Roof gardens, 57, 58, 59, 60, 61, 62, 149
Rotary washing lines, 100
Rusty pipes, 101

S

Scaffold boards, 101, 147
Sand pits, 101
Sanitary ware, 101
SalvoWeb, 125
Seaweed, 102
Shells, 102
Shoes, 102
Socks, 103
Soil Sacks, 103
Solar Power, 43
Sponges, 104
Steel plate sections, 103
Straw bales, 104
Stumpery, 21, 22, 23, 24, 25, 26, 27, 28
Sylvan Studio, 140

T

Tea bags, 104
Teapots, 104
Tights, 105
Timber, 105

Tools, 106, 116, 117, 118, 120, 139
Toothbrushes, 106
Tom Massey, 141
Tree management, 12, 106
Trugs, 106
Tyres, 106, 118, 149, 150, 153

U

Umbrellas, 107
Used compost, 108

V

Vertical Walls, 65, 66, 67
Victorian Kitchen Garden, 48, 51
Vintage objects, 109, 136, 138, 139

W

Wartime Kitchen Garden, 63, 64, 65
Washing machine rotary drums, 110, 143
Water storage, 18, 110
Wheelbarrows, 111, 118
Wheelie bins, 111
Windows, 111
Wire netting, 112
Wine bottles, 112
Wine boxes, 148, 149
Wooden crates, 113
Wormeries, 55, 56

Y

Yoghurt pots, 113